2025 NIGHT SKY ALMANAC

A Month-by-Month Guide to North America's Skies from The Royal Astronomical Society of Canada

Nicole Mortillaro

FIREFLY BOOKS

A FIREFLY BOOK

Published by Firefly Books Ltd. 2024
Copyright © 2024 Firefly Books Ltd.
Copyright © 2024 The Royal Astronomical Society of Canada
Text © 2024 Nicole Mortillaro
Photographs © as listed on page 128

First printing

ISBN: 978-0-2281-0470-4

Library of Congress Control Number: 2024935774

Published in Canada by
Firefly Books Ltd.
50 Staples Avenue, Unit 1
Richmond Hill, Ontario L4B 0A7

Published in the United States by
Firefly Books (U.S.) Inc.
P.O. Box 1338, Ellicott Station
Buffalo, New York 14205

Project manager/Editor (Firefly Books): Julie Takasaki
Project manager (RASC): Robyn Foret, Past President of The RASC
Technical editors: James S. Edgar FRASC (Editor, RASC *Observer's Handbook*)
    and Chris Vaughan
Graphic design: Noor Majeed and Stacey Cho
Illustrations and sky charts: Peter Kovalik

Printed in China | E

Canada

We acknowledge the financial support
of the Government of Canada.

Contents

Introduction

If you're reading this, you clearly love the night sky and all the joy it brings you.

This guide aims to provide novice and intermediate amateur astronomers with the knowledge they need to enjoy the night sky in 2025, from understanding planets, nebulae (clouds of gas and dust), comets, asteroids and meteors, to learning about annual and significant celestial events. This is your pocket guide to the cosmos.

But first, it's important that you understand how we navigate the night sky. Just like for navigation on Earth, astronomers use particular coordinates in the sky to figure out what we're looking at.

The **celestial sphere** is an imaginary sphere with Earth at its center. At any one time, an observer's night-sky view only includes half of this sphere, because the other half is below the horizon.

Earth's axis is tilted at 23.5 degrees to the **plane** of the Solar System – the plane being the orbit of the Earth around the Sun. For observers on the ground, the celestial sphere seems to rotate from east to west; that is why the Sun, for example, rises in the east and sets in the west.

The Earth's axis points less than 1 degree away from **Polaris**, the North Star, in the Northern Hemisphere. It's really important to know where north is, so you can navigate around the sky. Use a star chart, your smartphone or both to find true north at your observing location. (In the south, the Earth's axis points about one

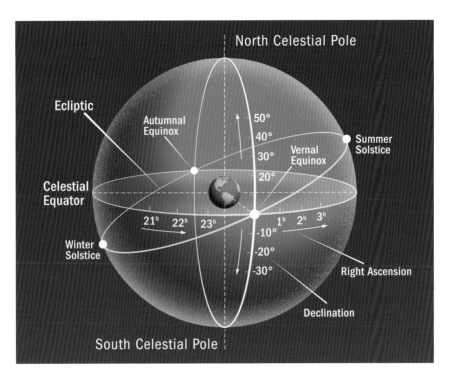

degree away from **Sigma Octantis**, a faint star; given Canada and the United States are in the Northern Hemisphere, however, we will focus on Polaris.) When you look at the sky over long periods of time, Polaris stays still as the other stars rotate around it.

You can easily see this phenomenon by setting up a camera, pointing it toward Polaris, and taking an extended (or long-exposure) photograph of the stars. You will see circular streaks in the sky as other stars rotate around Polaris. Some stars are so close to the North Celestial Pole that they never set. Other stars farther from the pole will rise in the east and set in the west, just like our Sun. Many stars are so far from the pole that they never rise above your horizon.

While beginner astronomers will navigate using a simplified star chart, it's helpful to know some of the terminology we use for finding our way around the sky. That way, as you gain confidence, you can follow professional astronomers in their observations.

The point directly overhead your observing location is called the **zenith**, and the **celestial equator** is directly above the Earth's equator. The point directly below you (under the horizon and opposite to the zenith) is the **nadir**. The line that runs from the north point on the horizon, through the zenith, to the south point on the horizon, is the **meridian**.

To navigate the night sky, astronomers created a type of latitude and longitude called **celestial coordinates**. In astronomy, we use **declination** (Dec.) and **right ascension** (RA). Declination is like latitude on Earth, running from north to south. Right ascension is like longitude on Earth, running from west to east. Declination is measured in degrees, and right ascension is measured in hours, minutes and seconds.

Star trails with Polaris at the center

As Earth rotates, the apparent position of most of the stars changes. More advanced astronomers may want to learn more about exactly where to find faint stars in their telescopes. For these situations, we navigate using **altitude** (angular elevation above the horizon, between 0 degrees at the horizon and 90 degrees at the zenith) and **azimuth** (the number of degrees clockwise from due north). But if you're just starting out, don't worry yet about these advanced observing techniques. A simple star chart will help you find the constellations and the planets.

If you want to look for planets, you should also pay attention to the **ecliptic**. That is the path the Sun takes through the constellations, and the Moon and planets do not stray far from that path.

Handy Sky Measures

Astronomers need to know how far apart things are in the sky. And they do this using angular degrees.

It's not hard to imagine the sky as a sphere that measures 360 degrees — after all, space surrounds Earth on all sides. Standing in one spot on Earth, if you trace the sky from horizon to horizon, that would equal 180 degrees. Remember, the other 180 degrees is under the horizon.

If you want to measure the distances between two objects — say, between the Moon and Venus as they appear together in the sky — you can use your hand as a measuring tool. It all lies within your fingers.

Hold your hand at arm's length. The width of your pinky finger equals 1 degree. The width of your three middle fingers held at arm's length equals five degrees; a closed fist is 10 degrees; the distance between the tip of your index finger and the tip of your pinky is 15 degrees; and the distance between your thumb and pinky is roughly 25 degrees.

This is particularly helpful when trying to see how high or low something is above the horizon.

You can practice measuring the degrees with stars found in the Big Dipper. The chart shows how to find the Big Dipper using Polaris. Because it's circumpolar, the orientation of the Big Dipper varies.

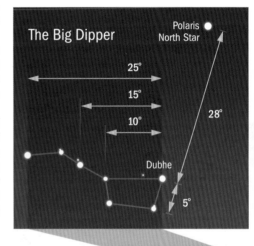

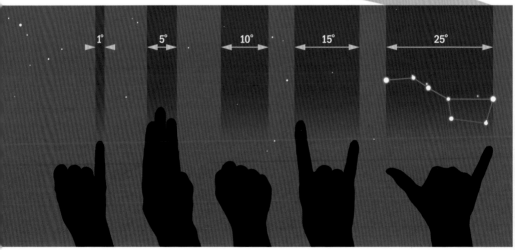

A view of Polaris and the Big Dipper over Dinosaur Provincial Park, Alberta

The Night Sky: A Cosmic Time Machine

Every time you look up at the sky, you are looking back in time.

Light from the Sun takes 8.3 minutes to reach us. Light from the Moon takes roughly 1.3 seconds. That's because light takes time to travel. In fact, light travels through space at roughly 300,000 kilometers (186,000 miles) per second.

We call the distance light travels in one year a **light-year**. So, when we look at stars, galaxies and nebulae, they appear how they looked back in time, depending on their distance to us. For example, the Andromeda Galaxy is the closest spiral galaxy to our own Milky Way. It is easily visible to the unaided eye in the Northern Hemisphere at dark-sky sites, even though it lies roughly 2.5 million light-years from us. That means, we are seeing the galaxy as it was 2.5 million years ago.

Sirius, also known as the "Dog Star," is the night sky's brightest star. It's visible during the winter months in the Northern Hemisphere. Its distance of 8.6 light-years from Earth means we are looking at it as it was 8.6 years ago.

Our Universe is roughly 13.8 billion years old. Astronomers improve their understanding of the evolution of our Universe by using powerful telescopes that can see farther and farther away, and so farther back in time.

Launched in 1990, the Hubble Space Telescope has provided us not only with jaw-dropping images of astronomical objects

Andromeda Galaxy (Messier 31), our closest galactic neighbor

Captured by NASA's James Webb Space Telescope and titled "Webb's First Deep Field," at the time this image was the deepest and sharpest infrared image of the distant Universe

such as galaxies and nebulae, but it has also helped narrow down the age of the Universe.

And more recently, in 2021, the James Webb Space Telescope opened its eyes for the first time, providing us with a look back to some of the very first galaxies to form — back to when the Universe was less than a billion years old.

It's almost like looking at a baby picture of our Universe.

While time machines may not be possible, every time we pick up a pair of binoculars or peer through our telescopes, we are taking a voyage back in time.

Binoculars and Telescopes

Using your eyes to move around the sky allows you to see the Moon, the planets, the Sun and even some objects outside of our Solar System. However, having binoculars or a telescope opens up the Universe to you.

Binoculars are an amazing first tool for seeking out the marvelous wonders of the night sky. A good pair of binoculars can reveal the intricacies of the Moon, showing its peaks and valleys, or uncover the daily motion of Jupiter's moons as they dance around our Solar System's largest planet. A decent telescope can make these objects appear bigger and show details of star clusters or nebulae.

But the question that often arises is what type of binoculars or telescope should I get?

For binoculars, it's important to understand two things: the **magnification** (or power) and the **aperture**. Binoculars are usually represented by two numbers, separated by an "×." Two example binocular numbers are 7×50 or 10×50. The magnification is the first number, and it represents the number of times larger something will appear compared to viewing it with the naked eye. The second number is the aperture in millimeters (about 3/64ths of an inch), and it represents the diameter of each lens. Good viewing in part depends on how well your binoculars can gather light. If your binoculars have a larger second number, they have a larger aperture and can gather more light from distant objects. But that comes with a cost. Bigger binoculars are heavier and, therefore, more difficult to hold, which means you will need a tripod to steady your view.

Another number often given for binoculars is the **field of view**, or FOV. This is how wide you will be able to see in degrees (see "Handy Sky Measures," on page 6). In general, the higher the magnification you use, the smaller your field of view will be. At times, this will mean you need to make a choice. Do you want to zoom in on a particular crater on the Moon, for example, or observe a more spread-out chain of mountains? Making these decisions is difficult for astronomers, too, so don't worry if it feels hard — it gets

A pair of binoculars is a great portable tool for observing the night sky

a bit easier to decide what to do with practice.

The top end of handheld binoculars is typically 7×50. Larger than that, it's best to get a tall tripod if you want the best view possible and to share views with others.

We recommend using binoculars for a few months before investing in a telescope. Once you are comfortable with binoculars, luckily, your knowledge will be useful because telescopes use many of the same definitions for observing.

Remember that light-gathering ability is important; a typical beginner's telescope usually ranges between 50 to 150 millimeters (2 to 6 inches) in **aperture**, the diameter of the main lens or mirror. The bigger the aperture, the more light the lens will gather, and the brighter the view will be. Invest in a sturdy tripod, too, so that the telescope will not shake when you use it.

The **magnification** of a telescope depends on the eyepiece used; magnification is calculated by dividing the focal length of the telescope by the focal length of the eyepiece (using like units). For example, a telescope with a focal length of 1,000 millimeters and a 10-millimeter eyepiece will magnify an object 100 times. The same 10-millimeter eyepiece in a 500-millimeter telescope would magnify an object 50 times.

While a bigger aperture and higher magnification may seem like the best way to go, it's best not to get too caught up in choosing a telescope with the highest numbers. What is important is how you plan to use it. Since most people are unlikely to have backyard observatories, the most important thing may be portability. Too heavy a telescope means you're unlikely to haul it out to the backyard or to a vacation spot, far from city lights.

When it comes to telescopes, there is a wide array of choices. Some of the most popular

A refractor telescope, which is best used for viewing planets, the Moon and double stars

are refractors, reflectors and compound telescopes. Each type has its own benefits, and it all depends on what you prefer. We therefore recommend you try testing out different telescopes at a local star party held by astronomy groups before making the investment. Alternatively, check reputable astronomy magazines or forums for recommendations for beginners; in most cases, all it takes is a little Internet searching and some patience.

A **refractor telescope** has a front lens that focuses light to form an image at the back, and an eyepiece that acts like a magnifying glass to allow your eye to focus on that image. Refractors tend to be more reliable, as their lenses are fixed in place and, therefore, don't get out of alignment as easily as some other types of telescope do. Refractors are best used for planetary and lunar observing as well as viewing double stars.

A **reflector telescope** uses a mirror to gather and focus light. The advantage for reflectors is they don't usually suffer from **chromatic aberrations**, when light of different wavelengths (i.e., colors) doesn't focus on the same point; this makes them ideal to observe distant objects like star clusters. Chromatic aberrations cause the different colors of the spectrum to split and the image to appear blurry, which isn't great when looking at a group of stars. A **Dobsonian telescope**, which is a variant of a reflector with a simple mount, gives you a far bigger aperture for far less than the cost of other telescopes.

Compound or **catadioptric telescopes** combine lenses and mirrors to form an image. The **Schmidt-Cassegrain**, a type of compound telescope, has a compact design that makes it quite popular. With this instrument, an astronomer can get a bigger aperture in a smaller sized telescope. This telescope type is also portable, making it easier to move the equipment into remote areas.

Star parties are great opportunities to check out different telescopes and chat with fellow enthusiasts

A reflector telescope (front) and a variant of a reflector telescope, a Dobsonian telescope (back), are ideal for viewing distant objects since they don't usually suffer from chromatic aberrations

A Schmidt-Cassegrain telescope, which uses both lenses and mirrors, is compact and best for viewing finer details on planets and the Moon

Stars

Stars come in many different varieties. Our Sun is a **yellow dwarf star**, on the **main sequence**, meaning that it's converting hydrogen into helium at its core, like most other stars. When a star does this conversion, it releases a tremendous amount of energy. This energy, in the form of sunlight, allows life to thrive on Earth.

The Sun, at 4.5 billion years old, is considered middle-aged. When it runs out of fusion fuel, five billion years from now, it will at first swell, becoming a **red giant** and engulfing the inner planets, then it will slough off its outer layers to become a **white dwarf**.

One of the most interesting types of stars, some might argue, are **red supergiants**. These colossal stars — roughly 1,400 times the mass of the Sun — have relatively short lifespans.

And when supergiants do stop fusion, they do so in a spectacular fashion, in an explosion called a **supernova**. A supernova occurs when a star can no longer convert hydrogen into helium; eventually the core is converted into iron. During a supernova, the star first collapses and then explodes outward, creating even heavier elements.

Betelgeuse, a star found in the left shoulder of Orion, is a red supergiant. While many far-away and faint supernovae have been witnessed in modern history in other galaxies, none have occurred in our own galaxy — the **Milky Way**. When Betelgeuse goes supernova, its brightness will rival the full Moon in our sky. If you're hoping to see Betelgeuse explode, you're not alone. However, estimates peg Betelgeuse's death for some time within the next 100,000 years.

The most common star in the Universe is a **red dwarf**, which is a cool star that's much smaller than the Sun. Astronomers often search for potentially habitable planets around these stars, because the inherent dimness and smaller size of red dwarfs makes it easier to spot planets. The stars, however, can be quite volatile, occasionally releasing a tremendous amount of radiation. Intense radiation is not a friendly process for most life forms.

The **Hertzsprung-Russell diagram** (H-R diagram) was developed in the early 1900s by Ejnar Hertzsprung and Henry Norris Russell, following the research findings by two Harvard computers, Annie Jump Cannon and Antonia Maury. The diagram plots the temperature of stars against their luminosity. Just as we do, stars go through certain stages in their lives.

Top 10 Brightest Stars in the Night Sky	
1.	Sirius
2.	Canopus
3.	Alpha Centauri A (Rigil Kentaurus)
4.	Arcturus
5.	Vega
6.	Capella
7.	Rigel
8.	Procyon
9.	Achernar
10.	Betelgeuse

The Harvard Observatory Computers

In 1881, Edward Charles Pickering, the director of the Harvard Observatory, hired a team of women to compute and catalog photographs of the night sky. Their work was incredibly important in providing the foundations of astronomical theory. Annie Jump Cannon, one of the computers, added to work done by fellow computer Antonia Maury and developed a system of classifying stars that is still used today.

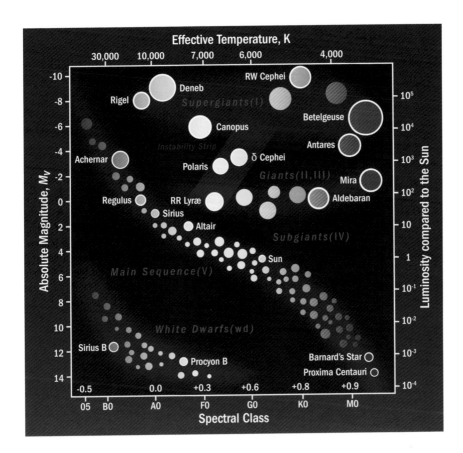

The H-R diagram provides astronomers with the information about a star's current age. Main-sequence stars that are fusing hydrogen into helium — such as our Sun — lie on the diagonal branch of the diagram.

Finally, there's a star's **magnitude**, or apparent brightness. Magnitude is measured on a scale where the higher the number, the fainter it is — and negative numbers are brighter than positive numbers. For example, Sirius, the brightest star in the night sky, measures -1.4 on this scale. Polaris is 2.0, and the Sun is -27. We also use magnitude to measure the brightness of other celestial objects, like the Moon, planets, asteroids and comets.

Note there is a difference between apparent and absolute magnitude. Apparent magnitude is the brightness of an object that we observe from Earth, but absolute magnitude is the brightness of an object if it were placed 32.6 light-years from Earth. This second measure helps astronomers directly compare the luminosity of objects and is what is used for the H-R diagram. On this scale, Sirius has a magnitude of 1.4, Polaris is -3.6 and the Sun is 4.8.

In astronomy, Greek letters of the alphabet are used to identify stars within a constellation usually from brighter to dimmer.

Constellations

Constellations, groups of stars that make imaginary images in the night sky, have been around since ancient times. Typically, the images astronomers use are based on Greek, Roman and Arabic mythologies. The International Astronomical Union (IAU) recognizes a total of 88 official constellations.

Some of the most recognizable constellations in the Northern Hemisphere are Orion (the Hunter), Cygnus (the Swan), Leo (the Lion), Gemini (the Twins), Scorpius (the Scorpion) and Ursa Major (the Great Bear).

More recently, there has been more effort to acknowledge constellations of Indigenous peoples. The naming of Indigenous constellations varies around the world, making these constellations regionally distinct. Some groups see Ursa Major as a bear, or a caribou. The Cree see Corona Borealis as seven birds and Cepheus as a turtle. To the Navajo, Polaris is Nahookos Bikq, meaning central fire. For some Indigenous peoples, such as the Inuit, the Northern Lights are dancing spirits.

Aside from the constellations, there are also **asterisms**, a group of stars within a constellation (or sometimes from several different constellations) that forms its own distinct pattern. The Big Dipper is probably the most famous asterism, as its stars lie within the constellation of Ursa Major. There's also the Summer Triangle, with bright stars from Cygnus, Lyra (the Lyre) and Aquila (the Eagle), and the Winter Triangle with stars from Orion, Canis Major (the Great Dog) and Canis Minor (the Little Dog).

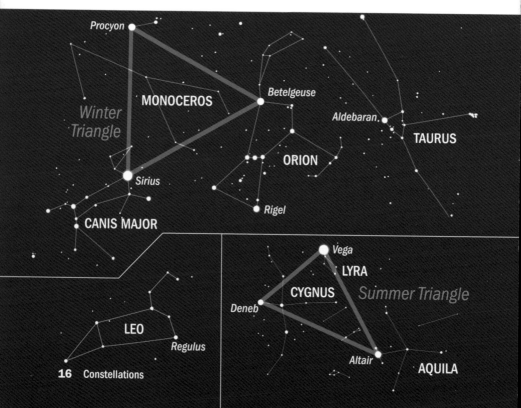

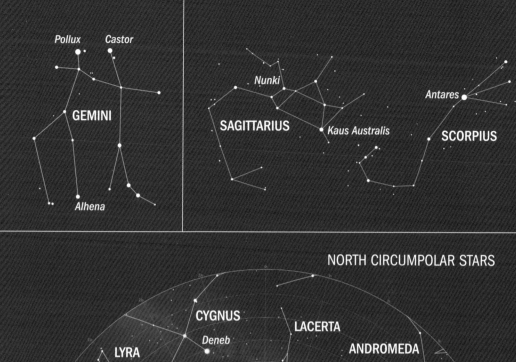

GEMINI

Pollux Castor

Alhena

SAGITTARIUS

Nunki

Kaus Australis

SCORPIUS

Antares

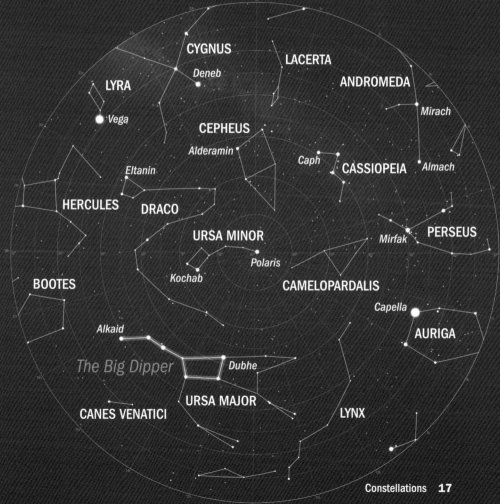

NORTH CIRCUMPOLAR STARS

CYGNUS

Deneb

LACERTA

ANDROMEDA

Mirach

LYRA

Vega

CEPHEUS

Alderamin

Caph

CASSIOPEIA

Almach

Eltanin

HERCULES

DRACO

URSA MINOR

Polaris

Mirfak

PERSEUS

Kochab

CAMELOPARDALIS

BOOTES

Capella

AURIGA

Alkaid

The Big Dipper

Dubhe

URSA MAJOR

LYNX

CANES VENATICI

Comets, Asteroids and Meteors

Comets are icy balls of debris — specifically, dust and ice — left over from the formation of our Solar System. They are sometimes referred to as "dirty snowballs" and can be stunning objects to see in the night sky when tails of dust and ionized gas fan out behind their cores. While we know of many comets, predicting the appearance of a bright one is all but impossible. Comets tend to fall apart before becoming too bright. The Sun's gravity and heat are strong, and comets themselves are very delicate visitors from the outer Solar System or beyond. Many comets literally crumble under the pressure as they dive in toward the Sun and the inner Solar System, where our planet resides.

Bright comets can be marvelous. On August 17, 2014, Comet Lovejoy C/2014 Q2 was spotted as it came toward the inner Solar System. This comet produced a spectacular tail as it neared the Sun. Tails occur when ice **sublimates**, turning directly from a solid into a gas.

There have been other wonderful naked-eye comets that have graced our night sky, including Comet Hyakutake C/1996 B2, Hale-Bopp C/1995 O1, and Comet NEOWISE C/2020 F3. Each comet has the year of its discovery in its official name, so NEOWISE was found in 2020, Hyakutake in 1996, and so on. Many Northern Hemisphere observers were treated to a special show when Comet NEOWISE appeared in the skies in 2020. It was one of the brightest comets to appear in a generation — even city dwellers could spot it through light pollution.

Periodic comets, or ones that we can predict, are given the designation "P," while those that appear unexpectedly are given the designation "C." For example, Halley's Comet, or 1P/Halley, appears roughly every 76 years — a clear P. The last time it passed was 1986; the next time will be in 2061. In general, C comets are brighter than P comets because P comets have sublimated their material into space from repeated trips by the Sun.

Like comets, **asteroids** (which are called minor planets) are left over from the formation of our Solar System. Instead of dust and ice,

Comet NEOWISE C/2020 F3

asteroids are rocks and come in all shapes and sizes. Most travel in the asteroid belt between Mars and Jupiter; however, they can be found other places, too, including beyond the orbit of dwarf planet Pluto at the outer edge of the Solar System, in a region called the Kuiper Belt. There are three broad composition types of asteroids:

Asteroid Bennu, taken by OSIRIS-REx

- **Chondrite (C-type)**, which are made up of clay and silicate rocks. They are dark in appearance and the most common. They are also the oldest type of asteroid.
- **Nickel-iron (M-type)**, which are believed to have experienced high temperatures and melted after they formed.
- **Stony (S-type)**, which are made up of silicates and nickel-iron.

Space agencies have sent spacecraft to collect samples from a few asteroids and return them to Earth. These samples can help us better understand the formation of our early Solar System. NASA's OSIRIS-REx spacecraft touched down on an asteroid called Bennu in October 2020 and left in May 2021 to deliver samples of Bennu to Earth. The samples arrived in September 2023, and soon afterward the spacecraft began a new mission as OSIRIS-APophis EXplorer (OSIRIS-APEX) and headed toward an asteroid called Apophis.

Like comets, asteroids can sometimes be knocked out of their orbits. In the asteroid belt, their orbit may be influenced by Jupiter's massive gravity. In the Kuiper Belt, they may be disturbed by interacting with other asteroids.

Asteroids known as centaurs travel in between the orbits of Jupiter and Neptune. Some of them may eventually be ejected from the Solar System, or travel past the inner planets (Mercury, Venus, Mars and Earth), or even burn up if they get too close to the Sun.

Asteroids that cross Earth's orbit are known as **potentially hazardous asteroids (PHAs)** or **near-Earth objects (NEOs)**. Asteroid impacts are a serious business. Sixty-six million years ago, an asteroid roughly 10 kilometers (6 miles) wide slammed into the area that is now the Yucatan Peninsula of Mexico. This event, sometimes referred to as the Cretaceous–Paleogene extinction, caused 75 percent of all animals on Earth to die out, including the dinosaurs.

While our planet does still exist in a shooting gallery (there are billions of objects in the Solar System, ranging from dust grains to kilometer-sized rocks), there are many organizations — including NASA and the European Space Agency — that are constantly searching for any objects that could cross Earth's orbit. To date, NASA says that it has identified nearly 90 percent of any objects 1 kilometer (0.6 miles) wide or larger. And the good news is that we currently know of no PHAs that are a threat to life on Earth up until the year 2200.

But that doesn't mean scientists aren't trying to prepare for the unexpected.

In the early morning hours of February 15, 2013, the sky over Chelyabinsk, Russia, became awash in light. A previously undetected object roughly 20 meters (66 feet) wide entered Earth's atmosphere and exploded in the sky, producing an airburst that shattered windows and injured more than 1,000 people.

Scientists are looking for ways to better detect these somewhat smaller objects, which may not be planet-destroying but could cause harm to populated areas. And they're even looking for ways to change their orbits.

In 2021, NASA conducted the first experiment trying to do just that. It launched its Double Asteroid Redirection Test, or DART. The small spacecraft traveled to a **binary asteroid** — a small asteroid called Dimorphos orbiting a larger asteroid called Didymos. The object of the test was to smash a spacecraft into Dimorphos to see if the impact could change its orbit. If humans can use a spacecraft to alter a PHA's orbit, it will no longer pose a threat to Earth.

The mission was a success. Before the impact, it took Dimorphos 11 hours and 55 minutes to orbit Didymos; after the impact, it took 11 hours and 23 minutes. Photographs by Hubble even showed that Dimorphos formed twin tails for some time after the impact (see the photo on the opposite page).

A **meteoroid** (from the Greek *meteoros*, meaning "high in the air") is a small piece of debris that has broken off from a larger object, usually an asteroid or a comet. Meteoroids are usually the size of a small pebble, but they can be smaller or a little larger. A **meteor** is the light, heat and (occasionally) sound phenomena produced when a meteoroid, collides with molecules in Earth's upper atmosphere. We also call meteors **shooting stars**.

When a meteoroid enters Earth's atmosphere, the surface of the object is heated. Then, at a height typically between 119.8 and 79.9 kilometers (74.5 miles and 49.7 miles), the meteoroid begins to **ablate**, or lose mass. Meteoroid ablation usually happens through vaporization, although some melting and

The vapor trail left after an object entered Earth's atmosphere over Chelyabinsk on February 15, 2013

breaking apart can also occur. If a meteoroid is large enough that it doesn't burn up entirely and reaches the ground, it is called a **meteorite**.

Meteoroids can be divided into two groups: **stream** and **sporadic meteoroids**. Stream meteoroids have orbits around the Sun that often can be linked to a parent object — in most cases a comet, but it can sometimes be an asteroid. When Earth travels within the orbit of a stream, we get a **meteor shower**. Because all the objects (meteoroids) move in the same direction, they all seem to come from one point in the sky, called the **radiant**. The constellation containing the radiant (or in some cases a nearby star) is what gives a meteor its name, such as the Perseids or Geminids (two of the most active meteor showers).

Sporadic meteors occur in random parts of the sky, with no radiant.

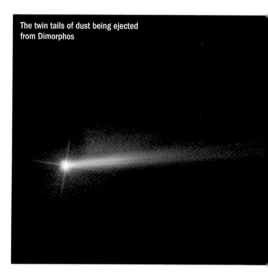
The twin tails of dust being ejected from Dimorphos

The Geminid meteor shower

Meteor Showers in 2025

There is nothing more wonderful than stepping out and looking up at the night sky, only to see a brief streak of light flash against the stars.

Almost monthly, we get major meteor showers. Some showers are best seen from the Northern Hemisphere, while some are better visible in the south, depending on the radiant.

When astronomers talk about the "peak" of a meteor shower, they're referring to the **Zenithal Hourly Rate**, or the ZHR. This is the rate of meteors a shower would produce per hour under clear, dark skies and with the radiant at the zenith. In practice, a single observer will likely see considerably fewer meteors than the ZHR suggests, owing to the presence of moonlight, light pollution, the location of the radiant, poor night vision and inattentiveness.

Here is a list of major meteor showers that you can enjoy simply by staying warm and looking up. No special equipment is needed beyond your eyes; just make sure to give yourself about 20 minutes to adjust to the darkness before searching for meteors.

Quadrantids:
December 27, 2024–January 10, 2025

The Quadrantids might be one of the best meteor showers of the year, with a ZHR of 120 for a brief interval of time. The only thing holding it back from earning the title is that January tends to be cloudy over North America, and the shower's peak has a brief window of six hours, making the average hourly rate closer to 25. But there's good news: the meteors often produce bright **fireballs**, or bright meteors with a magnitude brighter than −4.0. This shower's radiant lies between Boötes and Draco. This year there will only be a crescent Moon on the peak night, which won't interfere with the show.
Parent Object: Asteroid 2003 EH1
2025 Peak Night: January 3–4

Lyrids: April 16–30, 2025

After a dearth of meteor showers in the months of February and March, April brings us the Lyrids. This shower produces a ZHR of 18, so it's not a particularly strong shower, but the meteors do tend to produce fireballs. For the peak night, the Moon will be a thin waning crescent.
Parent Object: Comet C/1861 G1 (Thatcher)
2025 Peak Night: April 22–23

Eta Aquariids: April 19–May 28, 2025

The Eta Aquariids aren't a stellar show for the Northern Hemisphere, since the radiant rises in early dawn, but they can still produce a ZHR of 20 meteors, if you care to get up early enough. Rather than fireballs, this shower tends to produce long trains through the sky. The Moon will be at first quarter, which may prevent stargazers from seeing faint meteors.
Parent Object: Comet 1P/Halley
2025 Peak Night: May 5–6

Perseids: July 17–August 26, 2025

For those in the Northern Hemisphere, the Perseids are considered the best show of the year. The weather is warm and there are fewer clouds at this time of year. The shower's ZHR is 100, though rates of 50 to 75 are more commonly seen on the peak night. Unfortunately, there will be a waning gibbous Moon that will wash out all but the brightest meteors.
Parent Object: Comet 109P/Swift-Tuttle
2025 Peak Night: August 12–13

Orionids: October 2–November 7, 2025

This shower is considered medium strength, though it can sometimes surprise us with more activity. On average, the Orionids produce a ZHR of 10 to 20 meteors, though from 2006 to 2009, the shower produced roughly 50 to 75 an hour. The meteors that enter our atmosphere are fast — roughly 66 kilometers (41 miles)

per second — and, as a result, produce long, glowing trains. Fireballs are also possible. The great news is that there will be no Moon on the peak night.
Parent Object: Comet 1P/Halley
2025 Peak Night: October 21–22

Southern Taurids:
September 10–November 20, 2025
The Southern Taurids last for two months and have several minor peaks. Though this shower produces a ZHR of only 5, it can produce some fireballs. The shower is stronger in the Southern Hemisphere, though we can also catch a few meteors in the north. Unfortunately, there will be a full Moon on the peak night, meaning that only the brightest meteors will be visible.
Parent Object: Comet 2P/Encke
2025 Peak Night: November 5–6

Northern Taurids:
October 20–December 10, 2025
Like the Southern Taurids, the Northern Taurids last roughly two months and have a ZHR of 5. There are often reports of an increase of fireballs during the period when the two showers overlap. There will be a waning gibbous Moon on the peak night, which will interfere with this shower.
Parent Object: Comet 2P/Encke
2025 Peak Night: November 12–13

Leonids: November 6–30, 2025
While the Leonids don't produce a high number of meteors an hour (they have a ZHR of 15), they can produce **outbursts**, or meteor storms; the most recent such outbursts occurred in 1999 and 2001. Unfortunately, the next outburst isn't expected until 2099. Still, the Leonids do put on a show, with bright meteors and long-lasting trains. Stargazers will be pleased to know that there will be little interference by a thin crescent Moon.
Parent Object: Comet 55P/Tempel-Tuttle
2025 Peak Night: November 17–18

Geminids: December 4–17, 2025
Due to the weather — with increasing cloudiness and chilly temperatures — December isn't an ideal time to catch a meteor shower. But December is the month with the most active meteor shower of the year, called the Geminids. This shower has a ZHR of 150, and the meteors it produces are usually bright and colorful. There will be little interference from the Moon on the peak night, as it will be at last quarter.
Parent Object: Asteroid 3200 Phaethon
2025 Peak Night: December 13–14

Ursids: December 17–26, 2025
Right on the heels of the Geminids is the often-forgotten Ursid meteor shower. The ZHR for this shower is 10, but sometimes you might catch an outburst that could produce a ZHR of 25. The Moon will be a thin waxing crescent and will not interfere on the peak night.
Parent Object: Comet 8P/Tuttle
2025 Peak Night: December 22–23

The Moon

The Moon is most often the first astronomical object to capture anyone's attention in the night sky. It's also likely the first object most people see up close, whether it be through binoculars or a telescope.

Our nearest celestial neighbor has been an object of fascination since humans first looked up to the sky. Early scientists in ancient observatories carefully tracked the cycles of the Moon. Some 23 centuries ago, Aristarchus of Samos carefully observed a total lunar eclipse and derived impressively accurate measurements of the Moon's diameter. He estimated its diameter was one-third that of Earth. He was close: the actual percentage is 27.2, and its diameter is roughly 3,540 kilometers (2,200 miles).

It takes about 29.5 days for the Moon to go through its cycle of **phases**. When you can't see the Moon at all, it's called a **new Moon**.

First Quarter

Waxing Gibbous

Waxing Crescent

Full

New

Waning Gibbous

Waning Crescent

Third Quarter

When the Moon is fully illuminated, it is a **full Moon**. The list of phases is new Moon, waxing crescent Moon (when the Moon is just a thin crescent), first quarter Moon (when the eastern half is lit), waxing gibbous Moon (when the Moon is between half-full and full), and then full Moon. The process runs in reverse from full to new: waning gibbous Moon, third or last quarter Moon (lit on the western half), waning crescent Moon, and new Moon.

The Moon is tidally locked with Earth, meaning that we only ever get to see one face of it. The Sun does shine on the other side of the Moon when we can't see it, so don't call the far side of the Moon the dark side.

While it may seem that we always see the same Moon features month after month, that's not entirely true. The Moon oscillates from our perspective, a process called **lunar libration**. As a result, at certain times we see small zones along the Moon's edge that are normally hidden.

Observing the Moon

You don't need a telescope to enjoy the Moon. If observed even briefly on a regular basis, Earth's only natural satellite has much to offer the naked-eye observer. Keep an eye on it, and you will notice things like its wandering path through the constellations, the changing phases, frequent **conjunctions** (where two objects appear close together in the sky) with planets or bright stars, occasional eclipses, lunar libration, earthshine (illumination on the Moon reflected from our planet) and other wonderful atmospheric effects.

If you happen to take a look through a telescope, binoculars or even a camera, our closest astronomical target offers an amazing amount of interesting detail. There are countless features on the lunar nearside, and more than 1,000 have been formally named by the IAU. Many of the greatest names in astronomy, exploration and discovery are commemorated through these names. For extra dramatic effect, try looking at a lunar feature when the **terminator** — the line between darkness and sunlight — runs nearby the feature. The extra shadow could make it easier to see details in the object you're interested in looking at.

An estimated three to four billion years ago, our Solar System was bombarded by debris for hundreds of millions of years. This period is called the Late Heavy Bombardment, and you can see many of the scars from that time left over on the Moon.

Because the Moon has almost no atmosphere and no tectonic activity, those scars have remained over billions of years. Some of that activity created deep **impact basins** that filled with lava flows and hardened into basaltic rock. These largely circular regions were termed **maria**, or seas, by early astronomers because they resembled Earth's oceans.

The brighter white areas, which are highlands that surround the maria and dominate the southern portion of the Earth-facing hemisphere, feature many ancient **impact craters**.

After the bombardment slowed to a trickle, the Moon remained relatively free from intense impact activity — meaning the Moon has changed very little since billions of years ago. But, just like on Earth, meteorites still periodically slam into the Moon today.

The Moon map on the next two pages shows a few interesting features for you to target as you observe the Moon.

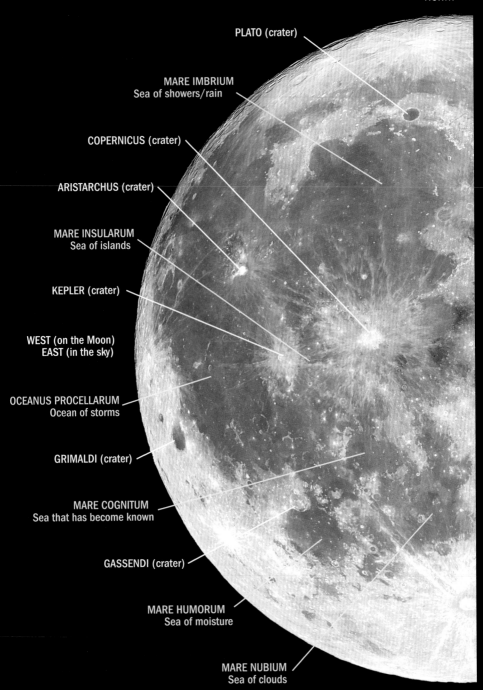

PLATO (crater)

MARE IMBRIUM
Sea of showers/rain

COPERNICUS (crater)

ARISTARCHUS (crater)

MARE INSULARUM
Sea of islands

KEPLER (crater)

WEST (on the Moon)
EAST (in the sky)

OCEANUS PROCELLARUM
Ocean of storms

GRIMALDI (crater)

MARE COGNITUM
Sea that has become known

GASSENDI (crater)

MARE HUMORUM
Sea of moisture

MARE NUBIUM
Sea of clouds

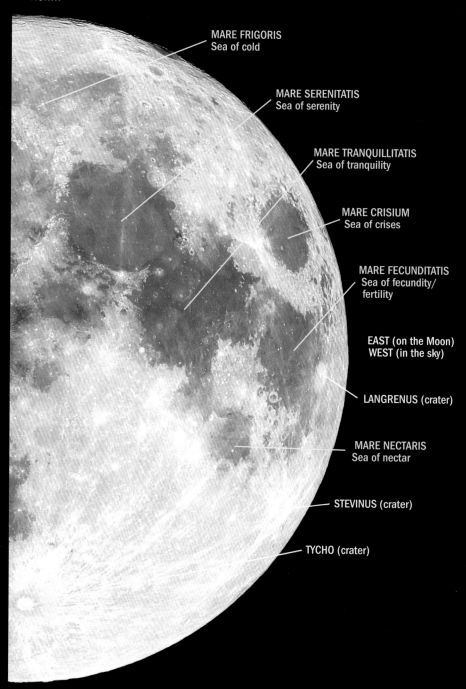

MARE FRIGORIS
Sea of cold

MARE SERENITATIS
Sea of serenity

MARE TRANQUILLITATIS
Sea of tranquility

MARE CRISIUM
Sea of crises

MARE FECUNDITATIS
Sea of fecundity/
fertility

EAST (on the Moon)
WEST (in the sky)

LANGRENUS (crater)

MARE NECTARIS
Sea of nectar

STEVINUS (crater)

TYCHO (crater)

SOUTH

The Sun

Our Sun highly influences Earth. It is connected with our seasons, weather, ocean currents and climate, and its power makes life itself possible.

Unlike Earth, the Sun isn't a solid body, and due to that, different parts of it rotate at different rates. At the equator, for example, it spins roughly once every 25 days, while the polar regions take about 30 days to complete a rotation.

In the core of the Sun, where hydrogen atoms fuse to make helium, temperatures are a searing 15 million degrees Celsius (27 million degrees Fahrenheit). The tremendous amount of energy generated at the core — **thermonuclear fusion** — is carried outward by radiation, taking roughly

Take Care of Your Eyesight

Do not observe the Sun without special protective equipment, such as a certified solar filter that covers your eyes or telescope aperture entirely. Unprotected observations even for a few seconds could damage your eyesight permanently.

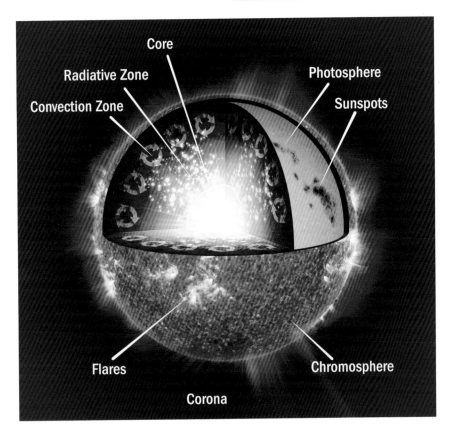

Core
Radiative Zone
Convection Zone
Photosphere
Sunspots
Flares
Chromosphere
Corona

170,000 years to get from the core to the top of the convective/convection zone. As hot as the core is, at the surface, it's a much different story — the temperature is a much "cooler" 5,500 degrees Celsius (10,000 degrees Fahrenheit).

The Sun also has a roughly 11-year cycle, that has a **maximum** (a period of increased solar activity) and a **minimum** (a period of low solar activity). During a maximum, the number of **sunspots**, which are cooler regions on the surface of the Sun, increases. Sometimes, the magnetic field lines of these sunspots can become entangled, finally snapping and releasing a tremendous amount of radiation into space. This process is called a **solar flare**. And often a **coronal mass ejection (CME)** can follow a flare, with charged particles of the Sun speeding outward into space. If these particles reach Earth, they can disrupt radio transmissions, damage satellites and, more positively, interact with our

A New Solar Cycle

At the end of 2020, we started Solar Cycle 25. After a quiet solar minimum, we can now expect to see more solar activity. Solar Cycle 24 was the weakest cycle in the last 200 years.

magnetic field. When the particles do interact with the field, they can produce an aurora.

Though beautiful, it must be said that these outbursts from the Sun can also cause power outages, as was experienced in Quebec in 1989. As such, astronomers are keen to better understand our nearest star using many spacecraft — such as the Parker Solar Probe, which was launched in 2018. Keep watching the eight-year adventure of the Parker Solar Probe as it studies the Sun's activity.

The Northern Lights

Observing the Sun

The Sun, our closest star, is a wonderful object to observe with any type of telescope, if used safely. The only safe filter is the kind that covers the full aperture of the telescope, at the front, allowing only 1/100,000th of the sunlight through the telescope. Without this filter, you risk permanently damaging your telescope or, far worse, your eyes. Ensure that the filter material is specifically certified as suitable for solar observing, and *take great caution every time you observe the Sun.*

The preferred solar filters are made of Thousand Oaks glass or Baader film. Thousand Oaks glass gives the Sun a golden-orange color, while Baader film gives the Sun a more natural white look, which can provide good contrast if there are any bright spots, or **plages**, on the Sun. Other solar-filter options include lightweight Mylar film (which gives the Sun a blue-white tint), and eyepiece projection. Projecting the image from the eyepiece onto white cardstock paper, or a wall, produces an image that can be safely shared with others.

While it might seem like the Sun is just this boring yellow ball in the sky, there are many things to see on its surface, including **prominences**, **filaments**, flares and sunspots. These phenomena result from the strong magnetism within the Sun, which erupts to the surface.

Prominences are solar plasma ejections, some of which fall back to the Sun in the form of teardrops or loops; against the Sun's disk, prominences viewed from above appear as dark filaments. Sunspots are cooler regions of the Sun that appear dark on the face of the surrounding surface. If the highly magnetic lines of sunspots intertwine, they can snap and cause a solar flare — a bright, sudden eruption of energy that can last from a few minutes to several hours. It should be noted that to see prominences and some other solar features, you need very specialized filters.

If you do not have the proper equipment to

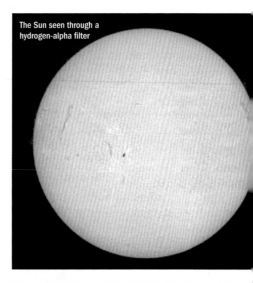

The Sun seen through a hydrogen-alpha filter

Sun with prominences, filaments, flares and sunspots

observe the Sun, you can always visit the Solar and Heliospheric Observatory (SOHO) and the Solar Dynamics Observatory (SDO) websites (see the list of resources on page 128) to see daily images of the Sun.

Eclipses

Eclipses — whether they're solar or lunar — are seemingly magical occurrences. In fact, that's exactly what many ancient civilizations believed. Several legends suggested a celestial being devoured the Sun during a total solar eclipse. For the Chinese, that being was either a dog or a dragon. The Vikings believed it was two wolves called Hati and Skoll. For the Vietnamese, it was a giant frog. Today, we understand that eclipses are awe-inspiring celestial events resulting from the movement and positions of the Earth, Moon and Sun. There can be as many as seven combined eclipses (i.e., solar and lunar) in a year, but no fewer than four.

A solar eclipse is truly a chance occurrence. The Sun's diameter is roughly 400 times that of the Moon, but the Sun is also about 400 times farther away from Earth than the Moon is from Earth. This means the Moon and the Sun appear roughly the same size in the sky, with only about 1/10th of a degree across of difference. In a **total solar eclipse**, the new Moon covers the entire face of the Sun, revealing the Sun's stunning corona and prominences. As totality begins and ends, sunlight peeking around the mountainous limb of the Moon creates fleeting effects such as the Diamond Ring and Baily's Beads. If the Moon were farther away or smaller, we wouldn't get this marvelous sight. In fact, the elliptical nature of both the Earth's orbit around the Sun and the Moon's orbit around the Earth means that the apparent diameters

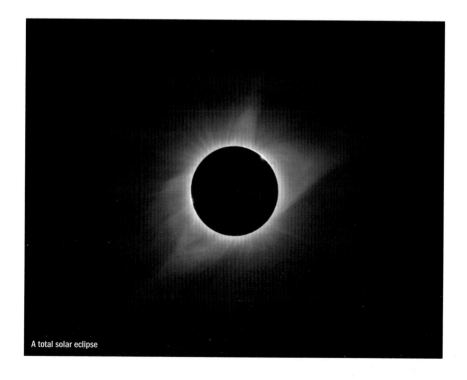

A total solar eclipse

of both objects vary significantly. Consequently, it is common for eclipse durations to vary by several minutes.

It should be noted that the totality only occurs in a very narrow path. Outside of this path, observers will only see a partial eclipse. That's why solar eclipse enthusiasts will travel great distances to take in the captivating sight of a totality. Remember, do not look at a solar eclipse unless you use special equipment recommended by The Royal Astronomical Society of Canada or the American Astronomical Society. The only time it is safe to look at a solar eclipse with the naked eye is when it is at its total phase, with the Sun completely covered by the Moon.

Because the Moon orbits at a 5-degree inclination to the ecliptic, we don't get a solar or lunar eclipse every month. Every year, there are two 36-day eclipse seasons during which either solar or lunar eclipses can occur, and the eclipse seasons move backward in time every year by 19 days. Also, because the Moon's orbit is elliptical, the angular size of the Moon varies along its path, and we sometimes have an **annular solar eclipse** (or "ring of fire"), when the Moon's disk does not completely cover the Sun's disk. There may be anywhere between zero to two total or annular solar eclipses in a given year. There may also be **partial solar eclipses**, when the Moon's central shadow entirely misses the Earth.

As for lunar eclipses, these occur when Earth is situated directly between the Sun and the full Moon. The Moon drifts through Earth's two shadows: the **penumbra**, which is the fainter, outer shadow, and the deeper shadow that is called the **umbra**. A **penumbral eclipse** is difficult to see with the naked eye as the brightness of the Moon doesn't appear to dim much. But a **partial** or **total lunar eclipse**, when the Moon passes through the umbra, is much more dramatic. During a total lunar eclipse, the Moon can turn an orange-red color, depending on Earth's atmosphere. There may be anywhere between zero to three total lunar eclipses in a given year. Typically, they occur about two weeks before or after a solar eclipse.

While not all eclipses will be visible from North America, you can watch online on sites like SLOOH or The Virtual Telescope Project. You can also see astronomer Fred Espenak's webpage eclipsewise.com for a comprehensive treatment of solar and lunar eclipses.

Eclipses in 2025

Last year's total solar eclipse that stretched across Mexico, the United States and Canada will be a hard act to follow. Unfortunately, North Americans won't get anything quite as spectacular as that, but the Moon and Sun will still put on a show across the planet, starting with a total lunar eclipse in March that will be visible in North America.

All the times listed are in UTC, or Coordinated Universal Time. See page 46 to learn how to calculate your local time from the time shown in UTC. Be sure to use a proper solar filter when

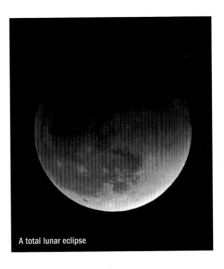

A total lunar eclipse

you are viewing a solar eclipse in person. And if you are clouded out, fortunately many of these events will also be live streamed online.

Lunar Eclipses

March 14, 2025: Total Lunar Eclipse

This year's eclipses start off with the first of two total lunar eclipses. This total lunar eclipse will be visible across North America, with most of the continent enjoying it in its entirety. The southern portions of the Moon should look noticeably darker than the northern. The Moon will be just 3.4 days from apogee (when the Moon is farthest from Earth in its monthly orbit).

Eclipse times (UTC)	
Partial eclipse begins (U1)	05:09
Greatest eclipse	06:58
Partial eclipse ends (U4)	08:48

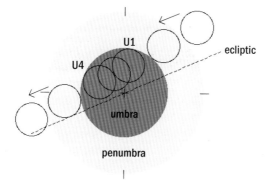

Total
2025 Mar 14
07:00 TD
06:58 UTC
Tot. = 65 m
Par. = 218 m
Gam. = 0.3485
U.Mag. = 1.1784
P.Mag. = 2.2595

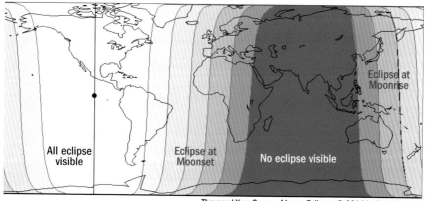

Thousand Year Canon of Lunar Eclipses © 2014 by Fred Espenak

September 7, 2025: Total Lunar Eclipse

This total lunar eclipse won't be visible from North America. The best views will be from Asia and western Australia, with parts of Europe and Africa seeing some of the event. This time, the northern parts of the Moon will be darkened more. This eclipse will occur when the Moon is 2.7 days from perigee (when the Moon is closest to Earth in its monthly orbit).

Eclipse times (UTC)	
Partial eclipse begins (U1)	16:27
Greatest eclipse	18:11
Partial eclipse ends (U4)	19:57

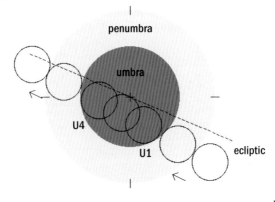

Total
2025 Sept 7
18:13 TD
18:11 UTC
Tot. = 82 m
Par. = 209 m
Gam. = -0.2752
U.Mag. = 1.3619
P.Mag. = 2.3440

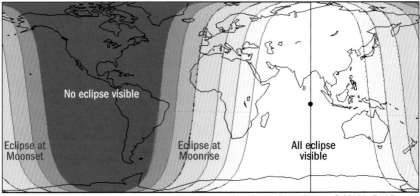

Thousand Year Canon of Lunar Eclipses © 2014 by Fred Espenak

Solar Eclipses

March 29, 2025: Partial Solar Eclipse

A partial eclipse occurs when the Moon only covers part of the Sun. Early risers in eastern and northern Canada will see much of the Sun covered by the Moon. Those in northwest Africa, Europe and northern Russia will see the eclipse in mid-day, with much less of the Sun eclipsed.

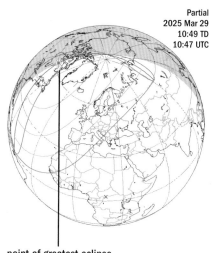

Partial
2025 Mar 29
10:49 TD
10:47 UTC

point of greatest eclipse

Thousand Year Canon of Solar Eclipses
© 2014 by Fred Espenak

September 21, 2025: Partial Solar Eclipse

This partial solar eclipse takes place as far south as you can get. It will be visible from the Southern Hemisphere across the South Pacific, New Zealand and Antarctica. It is considered to be a very deep partial solar eclipse, meaning that a very large part of the Sun will be blocked but only as viewed from near Antarctica.

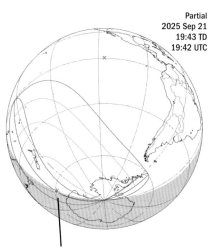

Partial
2025 Sep 21
19:43 TD
19:42 UTC

point of greatest eclipse

Thousand Year Canon of Solar Eclipses
© 2014 by Fred Espenak

The Northern Lights

The Sun appears to be a bright, unchanging orb in the sky. However, the Sun is anything but unchanging: As we know, it is a ball of constant activity, and that activity has a large influence here on Earth.

One such activity is a solar flare, which is often followed by a coronal mass ejection (CME) that sends particles speeding along the solar wind. Solar flares can reach Earth after a day or two, and they may disrupt radio transmissions. If Earth is in the path of a CME, the particles can travel down our magnetic field lines toward the poles; this creates the beautiful Northern and Southern Lights, or **aurora borealis** and **aurora australis**, respectively.

As mentioned earlier, the Sun goes through an average 11-year cycle of activity during which it experiences a solar minimum and a solar maximum. In the latter part of 2020, we began a new Solar Cycle, and as the cycle continues, we can expect to see more of the Northern Lights. It's worth noting that over the past few cycles the Sun has been less active than in the past.

Auroras come in different shapes and sizes. They can be steady, moving or rapidly pulsating. They also come in an array of colors, depending on how the particles interact with molecules at different altitudes. They are also hard to predict and can be difficult to see. Catching them is a special treat, even for experienced astronomers. Be aware that to the unaided eye the auroral colors are usually muted, as the light is not strong enough to stimulate our color vision. The bright colors seen in photographic reproductions of auroras show up when long exposures are used.

Green is the most common color and occurs when particles interact with oxygen molecules at an altitude of roughly 100 to 300 kilometers (62 to 186 miles). Between 300 to 400 kilometers (186 to 248 miles) the oxygen molecules produce a red color, and below 100 kilometers (62 miles) the particles interact with nitrogen — producing a pink color.

While the interaction of these solar particles can produce a beautiful light show, they can also be destructive.

One of the most powerful events from a CME was called the Carrington Event. In 1859, English astronomers Richard Carrington and Richard Hodgson were the first to ever witness a solar flare. But when the particles reached Earth, they disrupted telegraph systems in North America and Europe, in some reports even setting equipment on fire. The Northern Lights were visible as far south as Honolulu and the Southern Lights as far north as Santiago, Chile.

A massive CME, like the one that caused the Carrington Event, has a real chance of disrupting GPS and satellite communications and destroying electrical grids. As a result, space agencies have been launching more satellites and probes to monitor space weather, and power companies have been developing contingency plans to ensure the next powerful solar eruption doesn't cause such damage.

The Northern Lights over Saskatchewan

The Planets

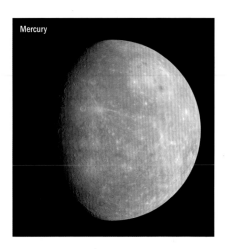

Mercury

As the planets (including Earth) revolve around the Sun, there are a number of notable events that take place, each of which has a specific name and meaning.

The inner planets Mercury and Venus never stray far from the Sun. When either planet is farthest from the Sun in the western evening sky, it is said to be at **greatest eastern elongation**; when either is farthest from the Sun in the eastern morning sky, it is said to be at **greatest western elongation**. These are the best opportunities to view the inner planets, though Venus can be seen at other times.

The term **conjunction** has a specific technical meaning — that is, when two objects have the same right ascension — but amateur astronomers also use the term casually when there is a close approach of two or more celestial objects.

Then there are **oppositions**: when the outer planets (Mars, Jupiter, Saturn, Uranus, Neptune) are opposite the Sun in the sky in right ascension. Around opposition dates, the outer planets appear closer, brighter and larger in diameter — great for telescopic viewing. The Sky Month-by-Month section (pages 46–119) lists eastern and western elongations, conjunctions and oppositions throughout 2025.

Mercury is a place of extremes. As the smallest of our Solar System's eight planets at one-third the size of Earth, Mercury is the closest planet to the Sun, sitting an average distance of 58 million kilometers (36 million miles) away. The planet, however, is only the second-hottest one in our Solar System. Mercury also orbits the Sun faster than any other planet and has the longest solar day, lasting 176 Earth days. Mercury has no moon.

The planet's surface warms to 427 degrees Celsius (801 degrees Fahrenheit) at the peak of its "noonday" heat; on the far side, its temperature drops down to a chilly –163 degrees

Celsius (–261 degrees Fahrenheit). Mercury's coldest locations are deep in the shadows of the craters, where the Mercury Surface, Space Environment, Geochemistry and Ranging (MESSENGER) spacecraft mission to the small planet discovered ice. Temperatures in crater shadows are at –183 degrees Celsius (–297 degrees Fahrenheit).

Due to the planet's proximity to the Sun, Mercury is a challenging target to observe. When it is visible in Earth's sky, it appears in a narrow window of time in the dawn twilight just before sunrise or in the evening twilight just after sunset, depending on whether it is west or east of the Sun in the sky. To observe Mercury, one needs a low horizon unobstructed by trees or buildings, as it is typically only 15 to 25 degrees away from the Sun when visible.

Venus, the brightest planet in the sky, is a spectacular sight to see. Known as both the "morning star" and the "evening star," Venus is similar in size to Earth. But it's definitely not a place you'd like to visit. The planet is covered in dense clouds and, as a result, suffers from a runaway greenhouse effect. The planet's surface has temperatures of 460 degrees Celsius (860 degrees Fahrenheit) almost constantly.

A view of the Martian landscape from NASA's Perseverance rover in March 2021

Venus spins in the opposite direction than the other planets, but very slowly. A day on Venus is longer than its year: one day takes 243 Earth days, while one year is 225 Earth days. Its orbit is, on average, 108 million kilometers (67 million miles) from the Sun. Being shrouded in white clouds, there is nothing much to see on Venus, but binoculars and small telescopes reveal that Venus passes through phases, similar to the Moon. (Mercury also shows phases, but they are more difficult to see, as the planet's disk appears so small.)

In 2025, Venus will be visible in the west after sunset from January to March. Mercury makes a brief appearance alongside Venus for the first few weeks of March before it is lost in the Sun's glare. By April, Venus transitions to a "morning star," visible in the east before sunrise.

There is no other planet that fascinates humans as much as **Mars**. In 1894, American astronomer Percival Lowell thought he could see deep canals carved through its surface that he believed were evidence of intelligent life. Later observations with spacecraft saw no evidence of these canals.

Mars is the most explored planet in our Solar System, and some even like to joke that it's the only planet with an entirely robotic "civilization" of Earth-made machines.

The orbiters, landers and rovers dispatched to the Red Planet have taught us a lot about its ancient past. Though rocky, dusty and seemingly devoid of any life, some evidence suggests Mars once had an ocean of water covering most of its northern hemisphere. Scientists sometimes cite the existence of water to suggest the planet was once habitable, though we have yet to find proof that life actually ever thrived there.

Mars is about half the size of Earth, with a day very close to Earth's 24-hour day. It lies roughly 228 million kilometers (142 million miles) away from the Sun. Small telescopes (with a 70 mm diameter and below) will show the red disk of Mars; a larger, good-quality telescope at high magnification will show surface features such as ice caps, dark basins and light deserts.

The mighty Mars begins 2025 in the eastern night sky and gradually rises higher as the months progress. During August, it will become hidden in the twilight after sunset.

Jupiter is the king of our planetary system, at 11 times wider than Earth. This gas giant boasts the most spectacular storm formation in our Solar System, called the Great Red Spot. Jupiter orbits 772 million kilometers (479 million miles) from the Sun and a single day is about 10 Earth hours long. It orbits the Sun in about 12 Earth years.

Jupiter is the second-brightest planet in our night sky and is also one of the most enjoyable planets to observe. With a modest telescope, you can easily see the cloud bands in its atmosphere. But you can even enjoy the planet through a pair of binoculars; if you watch the planet every night, four of its 95 confirmed moons – Io, Europa, Ganymede and Callisto – can be seen changing positions night after night.

The Great Red Spot (GRS) is a massive, swirling egg-shaped storm that has been shrinking for reasons that are unclear to astronomers. In the 1800s, the GRS was estimated at 41,000 kilometers (25,476 miles) along its long axis. NASA's Juno spacecraft measured the GRS at 16,350 kilometers (10,159 miles) in width on April 3, 2017. Several astronomy apps predict the best time on any night for viewing the GRS through a telescope.

Jupiter is high in the sky right from the beginning of 2025 and will be visible until the middle of May. It returns to the east in the early morning in July.

With its magnificent rings, **Saturn** is often considered the jewel of our Solar System. With a pair of binoculars, Saturn's pancake shape is evident. Look through a telescope, even a modest one, and its fine rings are clearly visible.

Saturn is nine times wider than Earth and 1.4 billion kilometers (886 million miles) away from the Sun. This gaseous planet has the second-shortest day in the Solar System, lasting roughly only 10.7 hours, while its orbit takes 29.4 Earth years. The planet also has more than 146 moons, though its largest, Titan, is the only one easily viewed through a modest telescope.

Saturn is the second-largest planet in the Solar System. It's not the only planet to have rings (Jupiter, Uranus and Neptune all have ring systems), but it has by far the most spectacular ring system we can see. Saturn's rings are made up of billions of pieces of ice that range from tiny dust-sized grains to chunks as big as a house. The ring system is roughly 282,000 kilometers (175,000 miles) across but only about 9 meters (30 feet) tall. There are several rings, mostly close to each other, but the main rings are called A, B, and C. The gap between the outer ring (A) and the first inner ring (B) is called the Cassini Division, following the discovery by Italian astronomer Giovanni Cassini.

Saturn will shine in the western evening sky until mid-February. When the planet reappears in the pre-dawn eastern sky during April, its rings will be razor-thin. Saturn will return to the evening sky from August onward.

Uranus is the faintest of the planets visible to the naked eye (in dark-sky conditions), and the third-largest planet overall. This giant ice planet has 13 faint rings and 28 small moons.

Jupiter

Saturn

But most interestingly, Uranus rotates on its side, at nearly a 90-degree angle.

In 1781, Uranus was the first planet to be discovered with the aid of a telescope by astronomer William Herschel. However, Herschel first believed it was either a star or a comet. It was confirmed as a planet two years later by Johann Elert Bode.

The planet is four times the size of Earth. A day on Uranus takes roughly 17 Earth hours, with one orbit taking 84 Earth years. Uranus orbits on average 2.9 billion kilometers (1.8 billion miles) from the Sun.

Uranus will be located near the Pleiades in Taurus during 2025. The planet will be brightest at opposition on November 21.

At an average distance of 4.5 billion kilometers (2.8 billion miles) from the Sun, **Neptune** is the farthest planet in our Solar System, taking about 165 Earth years to complete one orbit. Like Uranus, Neptune is roughly four times larger than Earth, and it, too, has a faint ring system.

Neptune also holds the distinction as being

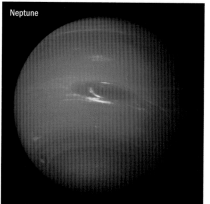

Neptune

the windiest planet in our Solar System, with clouds of frozen methane being whipped across the planet at 2,000 kilometers (1,200 miles) per hour. A day on Neptune is about 16 Earth hours long.

Neptune will spend 2025 near Saturn and reach peak visibility at opposition on September 23.

Deep-Sky Objects

The Universe has many amazing sights to behold, including stunning clusters of stars, nebulae and swirling galaxies.

There are two main types of star clusters: **open** and **globular**. Globular clusters are old star systems at the edge of spiral galaxies that can contain anywhere from thousands to millions of stars, packed in a close, roughly spherical form and held together by gravity.

Two beautiful globular clusters you can see from the Northern Hemisphere include Messier 13, found in the constellation Hercules (a hero from Greek mythology), and Messier 3 in the constellation Canes Venatici (the Hunting Dogs).

Open clusters are found on the **galactic plane**, the plane on which most of a galaxy's mass lies. They contain anything from a dozen to hundreds of stars, but the stars are more spread out than globular clusters. Perhaps the most famous open cluster, the Pleiades (Messier 45) can be spotted with the naked eye and through light pollution.

As well, there are **nebulae** — clouds of dust and gas. These are considered "stellar nurseries," as eventually the gas and dust will coalesce into new stars and potential stellar systems with planets, moons and, possibly, life. One of the most famous and easily visible is the Orion Nebula (Messier 42), found in the winter constellation Orion.

Some nebulae are leftover dust and gas from a **supernova**, an exploding star. There are also **emission nebulae**, which are clouds of interstellar gas excited by nearby stars that emit their own light at optical wavelengths. **Planetary nebulae** are cloudy remnants left over from stars that shed their gas and dust late in their lives. And finally, there are **dark nebulae**, interstellar clouds that are so dense they obscure the light of the objects behind them.

Messier 13, a globular cluster, seen here with Mars (at far left)

Messier Catalog

The Messier Catalog is a register of 110 objects in the night sky, including open and globular clusters, nebulae, galaxies and one double star. The catalog was started by Charles Messier in the 18th century. Messier was chiefly interested in finding comets, and the catalog is his list of non-comet objects that he observed. You can find a list of all the objects in the Messier Catalog on pages 120–125.

Pleiades (Messier 45)

Orion Nebula (Messier 42)

Galaxies

Galaxies come in many different shapes and sizes. There are four main types, however: elliptical, spiral, barred spiral and irregular. The diagram below shows Edwin Hubble's scheme for classifying galaxies, colloquially known as the "tuning fork" because of its shape.

Elliptical galaxies seem somewhat disorganized and look roughly egg-shaped. Shapes range from almost circular (E0) to very elliptical (E7).

Spiral galaxies have both a large central bulge and a thin disk of stars. Much like elliptical galaxies, these also range from tight spirals (Sa) to more diffuse (Sd).

Barred spiral galaxies are similar to spirals, but they have visible "arms" or bars near the center, and they range from tight (SBa) to diffuse (SBd).

There are also disk galaxies that don't have spiral arms. These are called **lenticular (lens-shaped) galaxies**. They are classified as S0.

The most common galaxy is the spiral, accounting for more than 75 percent of galaxies in the visible Universe. Our galaxy, the Milky Way, is believed to be a barred spiral. Our closest neighboring spiral galaxy is the Andromeda Galaxy (Messier 31), which is easily visible through binoculars or with the naked eye in dark skies.

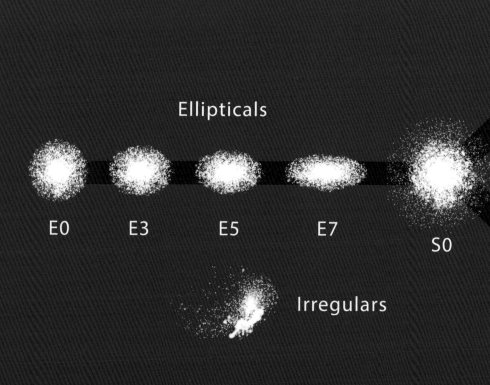

Ellipticals

E0   E3   E5   E7

S0

Irregulars

Andromeda Galaxy (Messier 31)

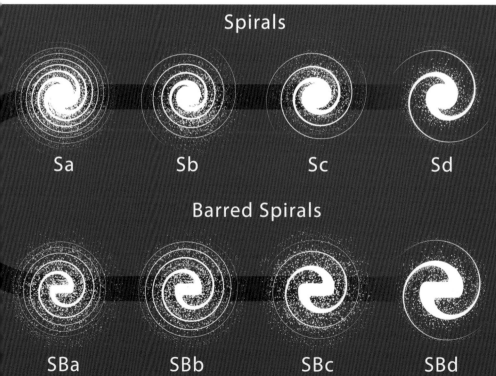

Spirals

Sa   Sb   Sc   Sd

Barred Spirals

SBa   SBb   SBc   SBd

The Sky Month-by-Month

Introduction

The following pages are your guide to the sky for every month in 2025. Each month features a calendar of events, information about Moon phases, sky charts facing south and north and descriptions of interesting objects to target.

Monthly Events

This section summarizes the events of each month, including Moon phases, conjunctions between the Moon and planets, conjunctions between planets and other planets, eastern and western elongations, meteor shower peaks and much more.

All times are listed in UTC, or Coordinated Universal Time, which is the standard time used by astronomers throughout the year. The map shown below will guide you on how to calculate your local time from the time shown in UTC. It's important to note that most of North America will observe Daylight Saving Time (DST) between March 9, 2025, and November 2, 2025. Between these two dates, you will need to add an additional hour to your local time.

You'll notice some of the times listed fall during daylight, when observation is likely impossible. However, you can use this date

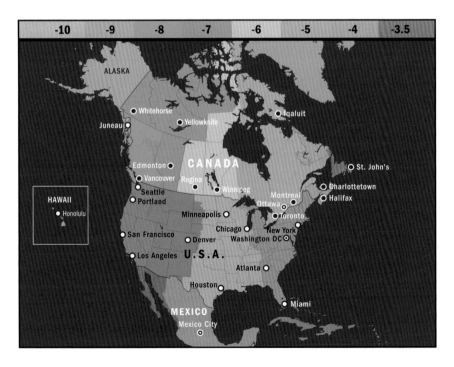

and time as a guide for observing an event on either the preceding night or the following night. Depending on your location, objects might also be below the horizon at the stated time, so again, use this calendar as a guide for which night these events can be observed. Similarly, the coordinates given as the distance between celestial objects may not reflect exactly what you will see in the night sky. This is particularly true for close approaches that take place during the day, local time. For example, the calendar of events might say Mercury is 1.5 degrees south of Jupiter at 21:00 UTC, but by the time Mercury and Jupiter become visible to an observer in Eastern Time living in Toronto, Canada, that separation might have increased to 2.4 degrees.

This section also features a calendar layout with the Moon phases for each day shown, as well as a description of interesting Moon events for the month. You might notice some discrepancies between the coordinates given in these descriptions and those listed in the table of events. Because the Moon moves in right ascension so quickly, conjunctions between the Moon and planets as listed in the table may appear different for North American observers, and the separation between the objects may be more than suggested. The coordinates in the descriptions have been altered to more closely reflect the view from North America, but of course there may be more variation based on when the objects are visible to you in the night sky.

The Moon phases shown in the calendar are based on the day they occur in UTC time. For some observers in North America, that might mean the ideal time to watch, for example, the full Moon rising would be the night before.

Sky Charts and Descriptions

Each month features two sky charts, one facing south and the other facing north. These charts highlight select constellations, stars, clusters, nebulae, planets and galaxies visible in the night sky. While these sky charts are an accurate representation of the sky, it should be noted that factors like light pollution, smoke, cloud cover and so on can affect how much you see, and some of the fainter stars shown on the charts may not be visible.

The charts are drawn at a latitude of 45 degrees north. Observers north or south of this latitude will see slightly more of the northern or southern sky, respectively. The charts show the sky at 10:00 p.m. local standard time on the 15th of each month. You will also have the same view of the sky at 11:00 p.m. local standard time at the beginning of each month and at 9:00 p.m. local standard time at the end of each month. We note these times on the sky charts and also include Daylight Saving Time in parentheses between March and November. The planets shown on the sky charts will move slightly relative to the stars over the course of the month.

The preceding month's charts can be used for sky viewing two hours earlier, and the following month's charts can be used for sky viewing two hours later. So, for example, if you wanted to view the sky at 8:00 p.m. local time in February, you would refer to January's sky chart. If you're referring to a previous or later month's sky chart, note that the positions of the planets will not be correct.

January

January's Events

We begin 2025 with the Quadrantid meteor shower. While it has the potential to be one of the most impressive of the year, with the ability to produce upward of 100 or more meteors an hour, it also has a short peak window of roughly six hours — compared with other showers that have peaks of two days. However, this shower does tend to produce bright fireballs. This year, the peak night is January 3 to 4, and the thin waxing crescent Moon that night will set early.

This month we also have four planets to enjoy. Jupiter will be parked high in the sky in the constellation Taurus while Mars will cruise past Pollux, the brightest star in the constellation Gemini, which will be climbing in the east. Uranus will be about 18 degrees from Jupiter, in Aries.

In the southwestern sky after sunset, Venus will be climbing while Saturn descends. On January 20, the pair will appear about 2 degrees apart — a great target for those with binoculars.

Calendar of Events

Day	Time (UTC)	Event
03	10:24	Venus 1.5°N of the Moon
03		Quadrantid meteor shower peak
04	2:59	Earth at perihelion: 0.9833 astronomical units (au)
04	12:18	Saturn 0.7°S of the Moon
06	23:56	First quarter of the Moon
07	18:35	Moon at perigee: 370,200 km (230,032 mi.)
09	20:01	Pleiades 0.3°S of the Moon
13	22:27	Full Moon
13	22:42	Mars 0.2°S of the Moon
14	16:03	Beehive 2.7°S of the Moon
15	20:17	Mars at opposition
18	10:53	Saturn 2.2°N of Venus
20	22:53	Spica 0.1°N of the Moon
20	23:55	Moon at apogee: 404,300 km (251,220 mi.)
21	20:31	Last quarter of the Moon
24	18:34	Antares 0.3°N of the Moon
29	12:36	New Moon
31	23:46	Saturn 1.1°S of the Moon

The Moon This Month

SUN	MON	TUES	WED	THURS	FRI	SAT
			1	2	3	4
5	6 1st Quarter	7	8	9	10	11
12	13 Full Moon	14	15	16	17	18
19	20	21 Last Quarter	22	23	24	25
26	27	28	29 New Moon	30	31	

On January 3, the 4-day-old Moon will meet with Venus in the west, separated by about 2 degrees. And the very next day, the Moon will pass close to Saturn, where they will be roughly 3 degrees apart.

On January 9, the Moon — which will be waxing gibbous — will be a sight to see as it crosses through the open cluster the Pleiades (Messier 45). On January 10, the Moon will reside near Jupiter for a beautiful pairing. On January 31, you can find a very thin crescent Moon near Saturn low in the west.

The full Moon rises on January 13, and the new Moon occurs on January 29. The Moon reaches perigee (the closest to Earth in its monthly orbit) on January 7 and apogee (the farthest from Earth in its monthly orbit) on January 20.

The Moon and Jupiter meet in the night sky

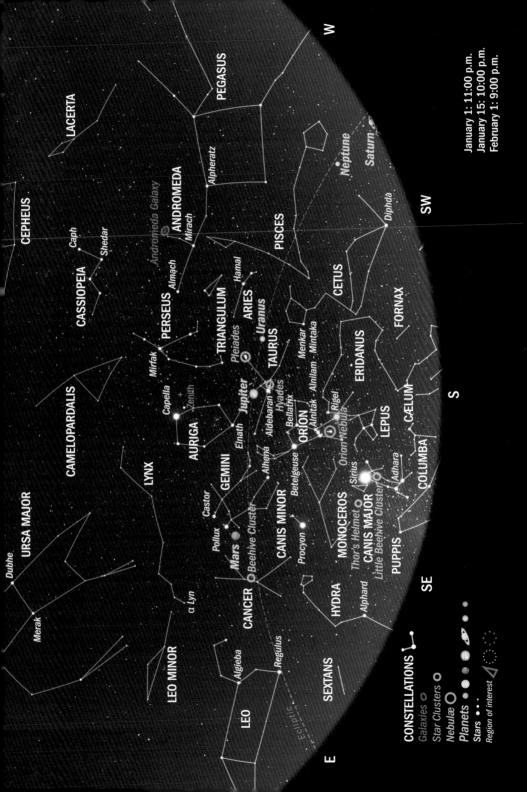

Highlights in the Southern Sky

While the nights are colder, they are also longer, which provides stargazers with a lot of dark hours to enjoy the night sky. And the best part is that there is an abundance of targets.

Beginning in the southeast, you can enjoy **Gemini (the Twins)** with its two brightest stars, **Castor** and **Pollux**. And close to Pollux is the unmistakable red planet, **Mars**. Just south of Gemini, you can find **Procyon**, the brightest star in **Canis Minor (the Little Dog)**. Procyon — the eighth brightest star in the night sky — is a double star that has a faint white dwarf companion.

And, speaking of bright stars, you can't miss **Sirius**, the brightest star in the night sky. It is part of **Canis Major (the Great Dog)**. Sirius is a binary star that lies 8.6 light-years away. It is roughly 25 times more luminous than our own Sun, part of the reason it appears so bright. No wonder the Greeks gave it a name that translates to "scorching" or "sparkling."

Using binoculars, you can find many star clusters in and around Canis Major, including the **Little Beehive Cluster (Messier 41)**, an open cluster that can be found just southwest of Sirius. There's also the beautiful nebula **Thor's Helmet (NGC 2359)**, which is a favorite photographic target for many astrophotographers.

But what takes the show is **Orion (the Hunter)**, a magnificent and unmistakable constellation in the southern sky. It has an abundance of jewels within it, including the most famous of all, the **Orion Nebula (Messier 42)**. This nebula is a rich star-forming region that is easily seen with the naked eye. In light-polluted cities, it's best viewed through binoculars. Within the heart of the nebula lies the **Trapezium**, an open cluster with four distinguishable stars that is best seen through a telescope.

Orion is most noted for its three stars that make up the belt of the Hunter, but also quite notable is **Betelgeuse**, a bright red star that makes up Orion's left shoulder (as seen from Earth). The red supergiant lies about 700 light-years away and is roughly 10 million years old, nearing the end of its life. The other bright star in Orion is **Rigel**, a blue supergiant that is roughly 870 light-years away. Rigel is more than 100,000 times more luminous than our Sun.

Thor's Helmet (NGC 2359)

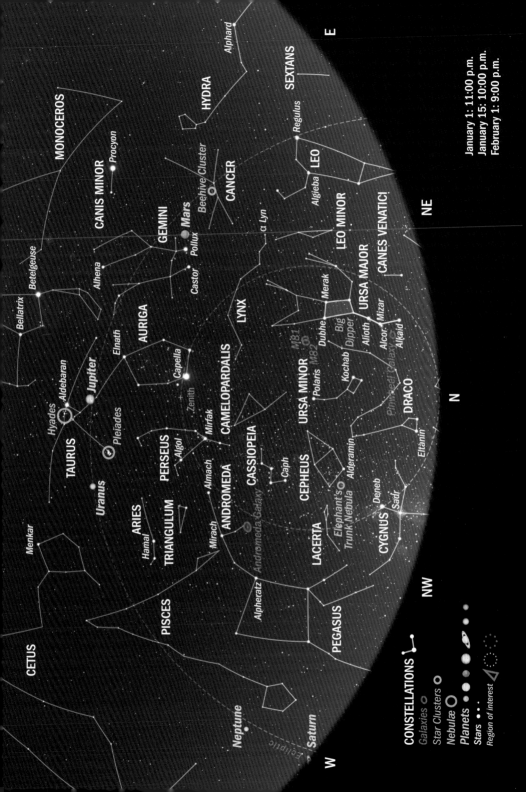

Pinwheel Galaxy (Messier 101)

Highlights in the Northern Sky

Low in the north, **Draco (the Dragon)** can be found slinking its way across the sky. While the head of the dragon, the star **Eltanin**, doesn't get very high above the horizon this month, the constellation is still impressive in that it is one of the longest in the sky. It is also a circumpolar constellation, meaning that it never sets below the horizon for many in the Northern Hemisphere.

Just north of Draco is **Ursa Minor (the Little Bear)**, home to **Polaris**, also known as the **North Star**. Polaris barely moves in relation to the rest of the stars in the sky: If you take a long exposure photograph with a camera pointed directly at Polaris, you would see the curved trails of the stars moving around it.

Ursa Major (the Great Bear) will be climbing the northeastern sky. It is home to one of the most famous asterisms, the **Big Dipper**, also known as the **Plough**. There are many notable stars here, beginning with **Alkaid** at the tip of the handle. The next star, **Mizar**, is a double star paired with the fainter **Alcor**. The pair can be seen unaided, but it makes for a great view through binoculars. The top two stars in the bowl are **Dubhe** (another double, though not as pronounced as Mizar and Alcor) and **Merak**.

Ursa Major is also home to the stunning **Pinwheel Galaxy (Messier 101)**, which sits near Alkaid.

Then there are **Bode's Galaxy (Messier 81)** and the **Cigar Galaxy (Messier 82)**, which lie near Dubhe. The pair is very close together and is best seen through any backyard telescope.

February

February's Events

The march of the planets continues in February.

Venus is well placed in the west after sunset. The Roman goddess's recent dance partner Saturn, however, will be sinking lower below it every night.

Mars will be shining between the bodies of the Gemini twins. On February 24, Mars ends its retrograde motion, meaning that it will start moving in the opposite direction, back toward Pollux. Meanwhile, Jupiter is still the king of planets in the sky, shining brightly in Taurus above the bright orange star Aldebaran. It is particularly well placed with the stunning Pleiades open cluster nearby.

Uranus is still in the constellation Aries.

The Gemini twins, Castor and Pollux, will be hosting Mars this month

Calendar of Events

Day	Time (UTC)	Event
01	15:27	Venus 2.4°N of the Moon
01	21:43	Moon at perigee: 367,500 km (228,354 mi.)
05	8:02	First quarter of the Moon
06	1:43	Pleiades 0.5°S of the Moon
07	3:00	Jupiter 5°S of the Moon
09	14:36	Mars 0.8°S of the Moon
11	0:03	Beehive 2.7°S of the Moon
12	13:53	Full Moon
17	20:11	Moon at apogee: 404,900 km (251,593 mi.)
20	17:32	Last quarter of the Moon
21	3:21	Antares 0.5°N of the Moon
28	0:45	New Moon

The Moon This Month

SUN	MON	TUES	WED	THURS	FRI	SAT
						1
2	3	4	5 1st Quarter	6	7	8
9	10	11	12 Full Moon	13	14	15
16	17	18	19	20 Last Quarter	21	22
23	24	25	26	27	28 New Moon	

On February 6, the Moon will be close to Jupiter and the Pleiades, this stunning open cluster

On February 1, Venus and the 3-day-old Moon will shine together in the southwest with Saturn below them.

On February 6, the waxing crescent Moon will perch above Jupiter and the Pleiades. And on February 9, the Moon will make another planetary visit – this time sitting near Mars in Gemini.

The Moon will be at first quarter on February 5 and will be full on February 12. The last quarter occurs on February 20, and the new Moon falls on February 28. It will be at perigee on February 1 and at apogee on February 17.

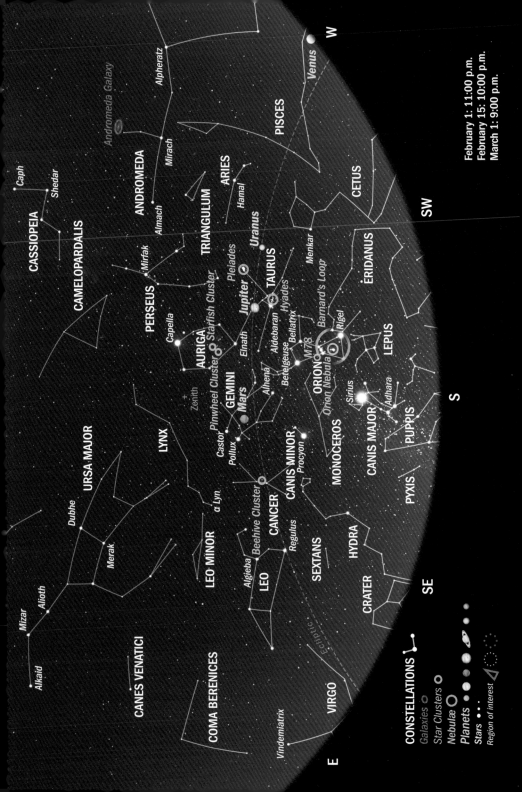

Highlights in the Southern Sky

Jupiter continues to dominate the southern sky when it comes to the planets, lying in **Taurus (the Bull)**.

Taurus is a wonderful constellation, home to the open cluster **Hyades (Melotte 25)**, which lies about 150 light-years from Earth. It makes an incredible sight in binoculars, with the star **Aldebaran** shining a bright orange-red. This star isn't actually part of the cluster itself. Instead, it is 65 light-years from Earth and just appears to be part of it.

Also in Taurus is the **Pleiades (Messier 45)**, another open cluster that is roughly 440 light-years from Earth. Also known as the **Seven Sisters**, the stars are hot, blue and very luminous. Even in moderately dark skies, they are visible. While we are most familiar with the seven brightest stars, there are about 1,000 in this cluster. And in photographs, the bright stars light up the surrounding stellar gas, making for truly stunning images.

Orion (the Hunter) is prominent in the south. This constellation is rich with targets, including **Messier 78**, which is located north of the leftmost belt star. This bright but diffuse nebula is 1,600 light-years away and is visible through small telescopes on dark nights.

One of the most stunning aspects in Orion is **Barnard's Loop**, a large, arc-like emission nebula that encompasses most of the constellation. Long-exposure photographs best reveal its immense structure. The **Orion Nebula (Messier 42)** is in the center of the arc.

Mars is still sitting nicely in **Gemini (the Twins)**. Southeast of Gemini is **Cancer (the Crab)**, where you can spot the bright **Beehive Cluster (Messier 44)** in binoculars.

North of Jupiter is **Auriga (the Charioteer)** with its brightest star, **Capella**. You can also find two open clusters within it using binoculars: the **Starfish Cluster (Messier 38)** and the **Pinwheel Cluster (Messier 36)**.

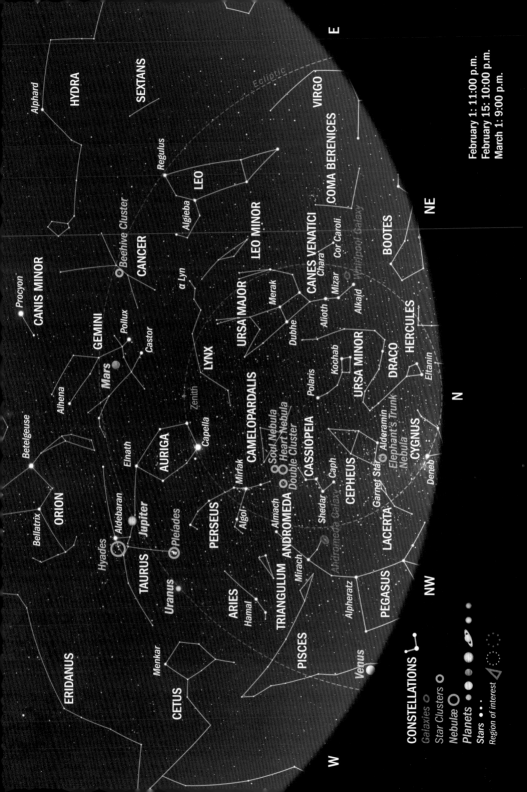

Heart Nebula (IC 1805) on the right and Soul Nebula (IC 1848) on the left

Highlights in the Northern Sky

The constellation of **Cepheus (the King)**, which is in the northwestern sky, looks like a house at this time of year. There are a few things that are notable about Cepheus, such as the **Garnet Star**, also known as **Mu (μ) Cephei**. It is found just beneath the square part of the constellation. This star stands out because of its bright red color – hence its name. It lies just north of the popular **Elephant's Trunk Nebula**, which is a favorite target for those who enjoy photographing the night sky.

Cassiopeia (the Queen) is higher in the northwest. It's known for its "W" or "M" formation, depending on where it is during the year. This month, it will be dangling from one end. Cassiopeia is a beautiful constellation to observe with binoculars, as it is abundant with stars and open clusters. Also within the constellation are two beautiful sights: the **Heart Nebula (IC 1805)** and the **Soul Nebula (IC 1848)**. The Heart Nebula is an emission nebula 7,500 light-years away and looks just like what you think it would – a heart. The pair span an area about 480 light-years across and are connected by a bridge of gas.

West of Cassiopeia you can find **Perseus (the Hero)**, with its bright star **Mirfak**. The constellation is the namesake of August's Perseid meteor shower due to where the meteors appear to be originating. This constellation is home to the **Double Cluster (NGC 869 and NGC 884)**, a pair of open clusters that appear close together in the sky.

Over in the northeastern sky, **Canes Venatici (the Hunting Dogs)** is somewhat of an unremarkable constellation, with just two stars, **Cor Caroli** and **Chara**. However, the region is a galaxy hunting ground for those with telescopes. Here you can spot the bright and well-known **Whirlpool Galaxy (Messier 51)** near the Big Dipper's handle.

March

March's Events

Every year in March, astronomers can challenge themselves to the Messier Marathon, attempting to see all 110 Messier objects in a single night. These nebulae, clusters and galaxies were recorded by the renowned French astronomer Charles Messier in the 18th century. Many of the objects can be found using binoculars, but many more can be seen through even small telescopes. Of course, weather can hamper any single observing night, so consider yourself lucky if you complete the challenge.

While Saturn has disappeared below the horizon, Mercury can now be found low in the western horizon just after sunset, with Venus nearby. If you like a challenge, try to find somewhere with a clear view of the west on March 13 or 14, when Mercury will be near enough on the left side of Venus for them to appear together in binoculars. But don't look for them until the Sun has fully set. Bright Jupiter will be joined by fainter Uranus in Taurus, and Mars will travel eastward to Pollux in Gemini.

In the wee hours of March 14, there will be a total lunar eclipse that will be visible from North America. The vernal equinox occurs on March 20, marking the beginning of spring.

Calendar of Events

Day	Time (UTC)	Event
01	16:18	Moon at perigee: 362,000 km (224,936 mi.)
05	2:30	Uranus 4.4°S of the Moon
05	12:15	Pleiades 0.6°N of the Moon
06	16:32	First quarter of the Moon
08	0:59	Mercury 18.2°E of the Sun (greatest elongation east)
09	0:50	Mars 1.4°S of the Moon
10	12:45	Beehive 1.9°S of the Moon
12	13:52	Venus 5.5°N of Mercury
14	6:55	Full Moon
14	6:58	Total lunar eclipse
17	12:37	Moon at apogee: 405,800 km (252,152 mi.)
20	5:02	Vernal (spring) equinox
20	17:58	Antares 0.5°N of the Moon
22	11:29	Last quarter of the Moon
29	10:47	Partial solar eclipse
29	10:58	New Moon
30	1:26	Moon at perigee: 358,100 km (222,513 mi.)

The Moon This Month

SUN	MON	TUES	WED	THURS	FRI	SAT
						1
2	3	4	5	6 1st Quarter	7	8
9	10	11	12	13	14 Full Moon	15
16	17	18	19	20	21	22 Last Quarter
23	24	25	26	27	28	29 New Moon
30	31					

We start the month with a beautiful conjunction of the Moon and Venus low in the west. On March 1, a faint, thin crescent Moon will be about 6 degrees away from the "evening star," and the two will be visible after sunset.

On March 5, the Moon will shine between Jupiter and the Pleiades, which should make for a beautiful sight. Binoculars will be ideal for viewing this. The next night, the Moon will still lie close to Jupiter.

The Moon makes a very close conjunction with Mars on March 8, when the pair will be separated by just 1.5 degrees.

The first quarter occurs on March 6, and the full Moon falls on March 14. The last quarter happens on March 22 and the new Moon on March 29. The Moon will be at perigee on March 1 and 30 and at apogee on March 17.

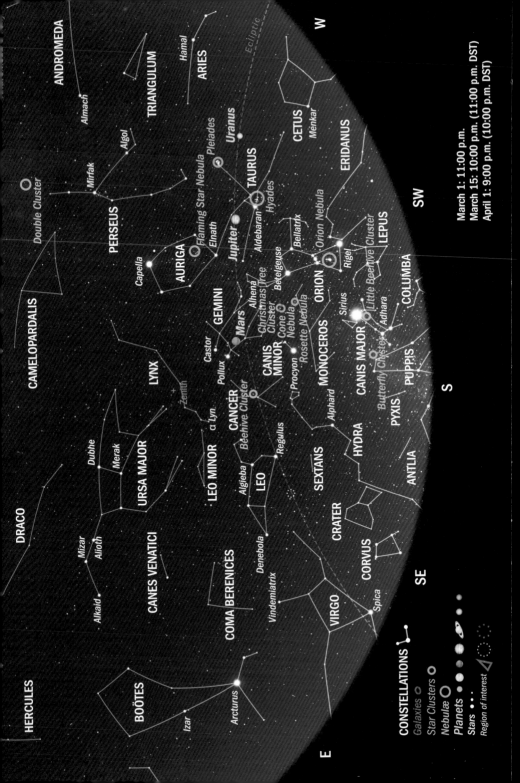

W

ANDROMEDA

Almach

Hamal

TRIANGULUM

ARIES

Algol

Mirfak

Double Cluster

PERSEUS

CETUS

Menkar

Ecliptic

Uranus

Flaming Star Nebula

Pleiades

Elnath

TAURUS

Jupiter

ERIDANUS

Aldebaran

Hyades

Capella

AURIGA

Bellatrix

Orion Nebula

LEPUS

Little Beehive Cluster

COLUMBA

Betelgeuse

Rigel

CAMELOPARDALIS

Alhena

Christmas Tree Cluster

GEMINI

Mars

ORION

Cone Nebula

Rosette Nebula

Castor

Sirius

MONOCEROS

Adhara

LYNX

Pollux

CANIS MINOR

CANIS MAJOR

Butterfly Cluster

PUPPIS

α Lyn.

Procyon

Zenith

PYXIS

CANCER

Beehive Cluster

S

Dubhe

LEO MINOR

Regulus

Merak

Alphard

ANTLIA

URSA MAJOR

Algieba

LEO

HYDRA

SEXTANS

Mizar

Alioth

Denebola

Alkaid

CANES VENATICI

Vindemiatrix

CRATER

DRACO

COMA BERENICES

CORVUS

VIRGO

HERCULES

Izar

BOÖTES

Spica

SW

Arcturus

SE

S

March 1: 11:00 p.m.
March 15: 10:00 p.m. (11:00 p.m. DST)
April 1: 9:00 p.m. (10:00 p.m. DST)

CONSTELLATIONS

Galaxies ○
Star Clusters ○
Nebulæ ○
Planets ● ● ● ·
Stars ● ● ·
Region of interest △

E

The Christmas Tree Cluster and Cone Nebula (NGC 2264)

Highlights in the Southern Sky

Leo (the Lion) can be found in the southeast. This is the constellation from which the Leonid meteor shower in November gets its name. Leo is marked by its brightest star, **Regulus** — one of the brightest stars in the sky. It is also the brightest in a four-star system and lies about 79 light-years from Earth.

Sirius, the brightest star in the sky, still sparkles in **Canis Major (the Great Dog)**. Nearby you'll also find the southern constellation of **Puppis (the Poop Deck)**. You can find a couple of very nice open clusters in this region of the sky. While these constellations do not climb very high above the horizon, if you are at a dark-sky site and have a clear line of sight to the southern horizon and binoculars, it won't disappoint.

Monoceros (the Unicorn) can be found to the left of **Orion (the Hunter)**. Monoceros is home to two beautiful targets. The spectacular star-forming **Rosette Nebula (NGC 2237)** is visible in binoculars and telescopes. The **Christmas Tree Cluster** and **Cone Nebula (NGC 2264)**, which is believed to contain upward of 2,500 stars, is about a palm's width above the Rosette Nebula. It is a very popular target for astrophotographers.

Another beautiful nebula is the **Flaming Star Nebula (IC 405)** found in **Auriga (the Charioteer)**. The nebula is about 5 light-years across and about 1,500 light-years away. It is best seen through medium-sized telescopes.

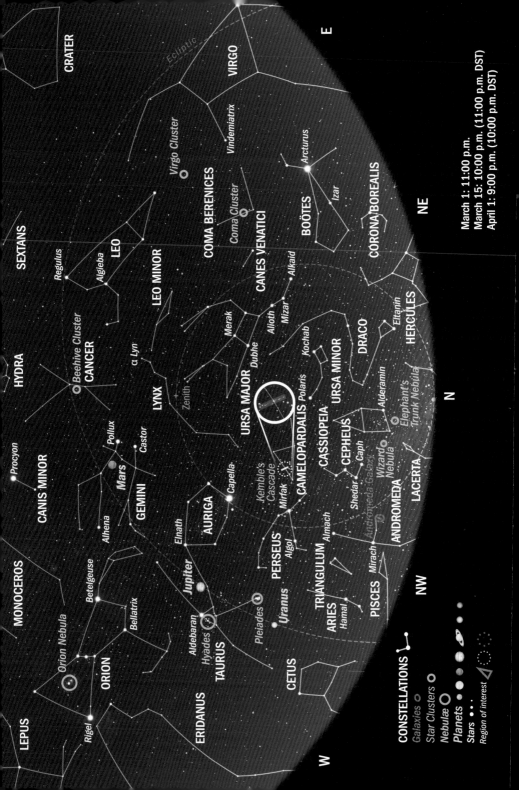

The asterism Kemble's Cascade

Highlights in the Northern Sky

Ursa Major (the Great Bear) is now riding high in the northeastern sky. The bear's tail — the star **Alkaid** — is pointing toward **Boötes (the Herdsman)**, which is creeping upward into the sky below it, marked by its brightest star, **Arcturus**.

Virgo (the Maiden) is also beginning to rise in the east, with **Coma Berenices (Berenice's Hair)** between Virgo and Boötes. What's notable about Coma Berenices is that it is home to the **Coma Cluster**, a cluster of thousands of galaxies. Also in this region lies the **Virgo Cluster**, which similarly contains thousands of galaxies.

Draco (the Dragon) is prominent in the sky, with the dragon's tail above Polaris in late evening.

Camelopardalis (the Giraffe) is just north of **Cassiopeia (the Queen)**. Camelopardalis is marked by its triangular shape. Though the constellation itself isn't very remarkable, it is home to **Kemble's Cascade.** That asterism of 20 somewhat faint stars arranged in a line stretches more than five Moon diameters across the sky and can be found with binoculars.

In **Cepheus (the King)**, you can find the beautiful **Wizard Nebula (NGC 7380)**, a star-forming emission nebula 7,000 light-years away that also contains an open cluster.

April

April's Events

April showers bring May flowers, but hopefully those showers only occur during the day. Astronomers need their clear night skies, especially now as the nights are shorter.

This month, there's a lot to enjoy in the night sky. Mars is still well placed, while Jupiter is beginning to sink in the west by the end of the month. If you have binoculars, see if you can spot Jupiter's four brightest moons — Io, Europa, Ganymede and Callisto. If you take a look through your binoculars the very next night, you'll notice that the moons have changed position. You can do that night after night to see just how quickly the Galilean moons orbit Jupiter.

If you're an early riser, you can find Mercury low on the eastern horizon just before sunrise, with Saturn to its right and bright Venus above them. On April 16, Mercury will be at its highest position in the morning sky, but it will be easier to see from the tropics. On April 30, Mars will begin an extremely close swing pass of the Beehive Cluster (Messier 44).

Last, but certainly not least, the Lyrid meteor shower peaks on the night of April 22 to 23. While the shower isn't well known for producing long trails, it can produce fireballs. Be sure to check out the night before the peak and the night after, as you may still catch a show.

Calendar of Events

Day	Time (UTC)	Event
01	16:28	Pleiades 0.6°S of the Moon
05	2:14	First quarter of the Moon
05	19:04	Mars 2.3°S of the Moon
06	16:44	Beehive 2.8°S of the Moon
10	8:19	Saturn 2.1°SW of Mercury
13	0:22	Full Moon
13	18:48	Moon at apogee: 406,300 km (252,463 mi.)
16	22:19	Antares 0.4°N of the Moon
21	1:35	Last quarter of the Moon
21	14:59	Mercury 27.4°W of the Sun (greatest elongation west)
22		Lyrid meteor shower peak
25	4:15	Saturn 2.3°S of the Moon
26	1:05	Mercury 4.3°S of the Moon
27	12:15	Moon at perigee: 357,100 km (221,892 mi.)
27	19:31	New Moon
29	6:35	Pleiades 0.5°N of the Moon

The Moon This Month

SUN	MON	TUES	WED	THURS	FRI	SAT
		1	2	3	4	5 1st Quarter
6	7	8	9	10	11	12
13 Full Moon	14	15	16	17	18	19
20	21 Last Quarter	22	23	24	25	26
27 New Moon	28	29	30			

It's no April Fool's joke: On April 1, the thin, waxing-crescent Moon will be sitting beside the Pleiades, which will provide a great target for binoculars. The very next night, you can see the Moon and Jupiter relatively close together in Taurus. On April 5, the Moon swings by Mars, which will be sitting in Gemini. On April 25, you can find Mercury, the thin crescent Moon and Venus close together in the early morning sky, just before sunrise. Saturn will be just south of Venus.

The Moon will be at first quarter on April 5 and will be full on April 13. Last quarter falls on April 21, and the new Moon is on April 27. Finally, the Moon will be at apogee on April 13, giving us the smallest full Moon of 2025, and at perigee on April 27.

Earthshine on the Moon

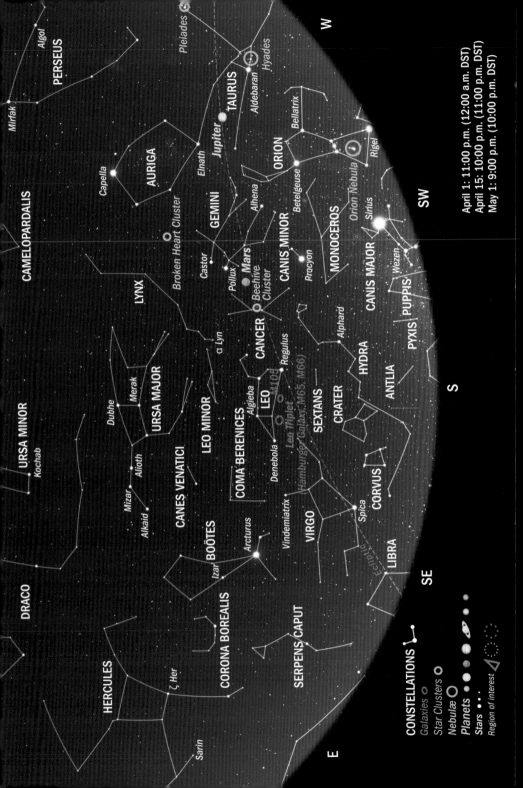

The Leo Triplet, consisting of the Hamburger Galaxy (NGC 3628), Messier 65 and Messier 66.

Highlights in the Southern Sky

While not a prominent constellation, **Hydra (the Water Snake)** is slithering its way across the southern sky. The constellation winds its way beneath **Corvus (the Crow)** and **Crater (the Cup)**, both of which are relatively small constellations.

Leo (the Lion) is very well placed in the southern sky. While it is host to a few galaxies, including **Messier 105**, it is best known for the **Leo Triplet**, which consists of the Hamburger Galaxy **(NGC 3628)**, **Messier 65** and **Messier 66**. The trio can be seen in a single view through even modest telescopes in a dark sky.

Cancer (the Crab) is also high in the south. **Mars** will begin the month near **Pollux** in **Gemini (the Twins)**. Then the Red Planet approaches the **Beehive Cluster (Messier 44)** in Cancer near the end of the month.

Orion (the Hunter) is now sinking in the west. You can still see **Auriga (the Charioteer)** with its most prominent star, **Capella**. Through binoculars, you can find a few nice open clusters near Auriga, including the **Broken Heart Cluster (NGC 2281)**, which can be found to the west of **Castor** in Gemini.

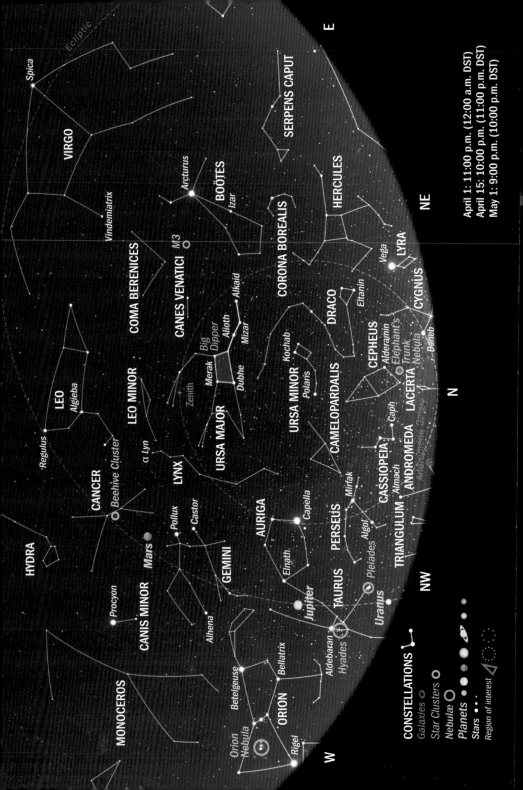

E

Spica

VIRGO

Ecliptic

Arcturus

SERPENS CAPUT

Vindemiatrix

BOÖTES

Izar

HERCULES

NE

M3

COMA BERENICES

CANES VENATICI

Alkaid

CORONA BOREALIS

April 1: 11:00 p.m. (12:00 a.m. DST)
April 15: 10:00 p.m. (11:00 p.m. DST)
May 1: 9:00 p.m. (10:00 p.m. DST)

Vega

LYRA

Regulus

LEO

Algieba

LEO MINOR

Big
Dipper

Alioth

Mizar

Eltanin

CYGNUS

Merak

Zenith

Dubhe

DRACO

Deneb

CANCER

Beehive Cluster

α Lyn

URSA MAJOR

Kochab

URSA MINOR

Polaris

CEPHEUS

Alderamin

Elephant's
Trunk
Nebula

N

HYDRA

Pollux

Castor

LYNX

CAMELOPARDALIS

LACERTA

Caph

ANDROMEDA

Andromeda Galaxy

Procyon

CANIS MINOR

Mars

Alhena

GEMINI

AURIGA

Capella

Mirfak

PERSEUS

CASSIOPEIA

Almach

TRIANGULUM

NW

Elnath

Algol

Pleiades

Urapus

Jupiter

TAURUS

Aldebaran

Hyades

Bellatrix

Betelgeuse

ORION

Orion
Nebula

Rigel

MONOCEROS

W

CONSTELLATIONS

Galaxies ○
Star Clusters ○
Nebulæ ○
Planets
Stars • · ·
Region of interest

Messier 3

Highlights in the Northern Sky

Ursa Major (the Great Bear) is now high up in the north. The spoon of the **Big Dipper** — the asterism within the constellation — is almost upside down. The curving bear's tail, tipped by the star **Alkaid**, points toward **Boötes (the Herdsman)**. Its brightest star is **Arcturus**, which is also the fourth brightest star in the sky. It lies about 37 light-years from Earth. Interestingly, though we can see the Southern Hemisphere star **Sirius** during the winter months in the Northern Hemisphere, magnitude –0.05 Arcturus is the brightest star north of the celestial equator.

Arcturus can also lead you to the beautiful globular cluster **Messier 3**, which can be found in the small constellation **Canes Venatici (the Hunting Dogs)**. It is one of the largest globular clusters in the sky, containing upward of half a million stars and sitting roughly 34,000 light-years from Earth.

Meanwhile, the constellation **Hercules** can be found low in the northeast, ready to begin its upward trek into the spring night sky. Nearby, **Draco (the Dragon)** can be found wrapping itself around **Ursa Minor (the Little Bear)** and its pole star **Polaris (the North Star)**.

May

May's Events

It's now warmer out, which is good news for astronomers, even though the darkness doesn't last as long as during the winter months.

Mars will travel through Cancer for most of May, while Jupiter disappears into the western twilight. Over the nights of May 3 to 5, be sure to pull out those binoculars for a beautiful pairing of Mars with the Beehive Cluster (Messier 44). The Red Planet will be so close that it will appear as part of the cluster and stand out nicely. The brightest of all the planets in the sky, Venus, will be a "morning star" that's visible before sunrise, with Saturn just to the southwest.

As we begin to transition to the summer constellations, both Scorpius and Sagittarius begin to appear low in the southeastern sky in late evening.

Finally, it's time for the Eta Aquariid meteor shower peak. The shower begins in early April, but peaks on the night of May 5 to 6. Although it is best viewed from the Southern Hemisphere, those in the Northern Hemisphere can expect to see roughly 10 to 30 meteors an hour under dark skies, especially before dawn after the bright Moon has set.

Calendar of Events

Day	Time (UTC)	Event
04	0:12	Mars 2.2°S of the Moon
04	0:27	Beehive 2.6°S of the Moon
04	13:52	First quarter of the Moon
04	23:59	Beehive 0.4°N of Mars
05		Eta Aquariid meteor shower peak
10	20:49	Moon at apogee: 406,200 km (252,401 mi.)
12	16:56	Full Moon
14	4:10	Antares 0.3°N of the Moon
20	11:59	Last quarter of the Moon
22	17:51	Saturn 2.8°S of the Moon
23	23:52	Venus 4°S of the Moon
25	21:37	Moon at perigee: 359,000 km (223,072 mi.)
27	3:02	New Moon
31	8:00	Beehive 2.3°S of the Moon

The Moon This Month

SUN	MON	TUES	WED	THURS	FRI	SAT
				1	2	3
4 1st Quarter	5	6	7	8	9	10
11	12 Full Moon	13	14	15	16	17
18	19	20 Last Quarter	21	22	23	24
25	26	27 New Moon	28	29	30	31

On May 3, grab your binoculars to see the waxing crescent Moon near both Mars and the Beehive Cluster.

If you're an early riser, on May 22, you can find the waning crescent Moon paired with Saturn in the early morning sky, just before sunrise. The ringed planet is best seen through binoculars, which might show you blue Neptune just to Saturn's left. Then, on May 23 to 24, there will be a conjunction of the Moon and bright Venus, which is unmistakable just before sunrise.

On May 31, the waxing crescent Moon will be roughly 3 degrees from Mars in the southwestern evening sky.

The Moon will be at apogee on May 10 and at perigee on May 25. The first quarter of the Moon is on May 4. The full Moon brightens the sky on May 12, while the last quarter is on May 20. And finally, the new Moon will fall on May 27.

On May 3 the Moon will be near Mars and the beautiful Beehive Cluster (Messier 44)

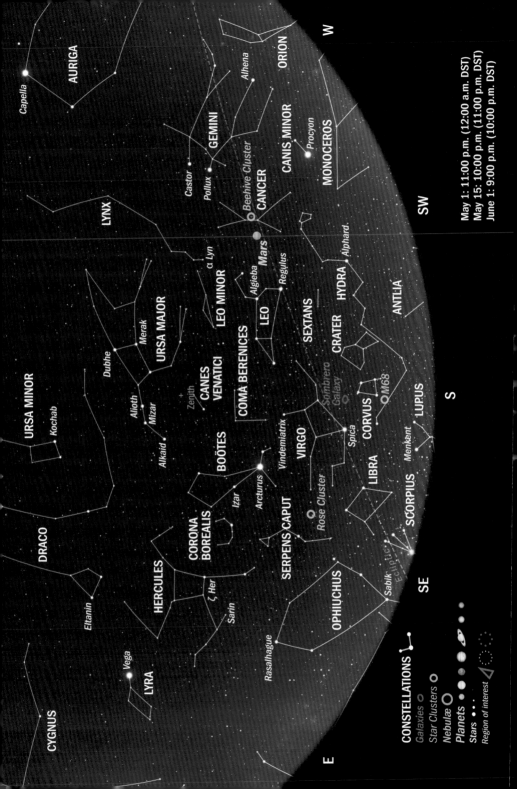

Highlights in the Southern Sky

Mars can be found in **Cancer (the Crab)** shining brightly, though it's somewhat dimmer than it has appeared over the past few months.

You can also find **Gemini (the Twins)** setting in the west, with its two brightest stars, **Pollux** and **Castor**, quite prominent in the sky, even in light-polluted cities. Southerly observers can see **Sirius**, found in **Canis Major (the Great Dog)**. The bright star is also dipping low in the west but is unmistakable.

To the south, you can find **Hydra (the Water Snake)**, wrapping itself across the sky. The constellation isn't particularly brilliant; however, it is home to some great objects, including the globular cluster **Messier 68** near the snake's tail. This cluster is best seen from dark-sky locations with binoculars or through medium-to-large telescopes.

Virgo (the Maiden) is well placed in the south. Straddling her border with **Corvus (the Crow)** is the magnificent **Sombrero Galaxy (Messier 104)**. This galaxy, which lies roughly 29 million light-years away, is an awe-inspiring sight. Zones of dark interstellar dust can surround galaxies that we see edge-on. The Sombrero is best admired through at least small telescopes, but medium-sized telescopes will reveal its prominent dust lane.

Serpens Caput (the Head of the Serpent) lies in the southeast. One of the most notable objects in this constellation is the **Rose Cluster (Messier 5)**, a bright globular cluster that is best seen using binoculars or a small telescope. It is about 24,500 light-years away and contains at least 100,000 stars, most of which formed 12 billion years ago.

South of Serpens Caput, you can find **Libra (the Scales)** rising in the southeast.

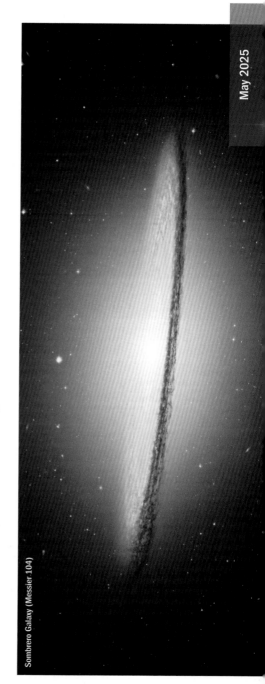

Sombrero Galaxy (Messier 104)

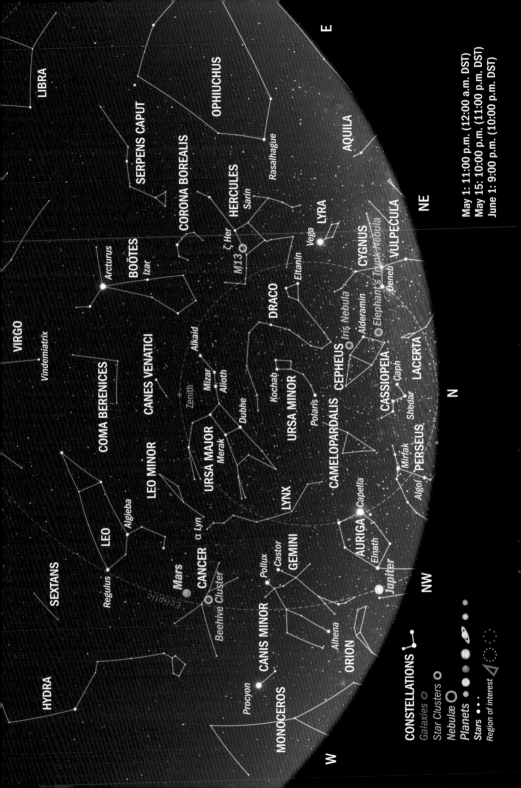

E

LIBRA

OPHIUCHUS

SERPENS CAPUT

CORONA BOREALIS

HERCULES

Rasalhague

AQUILA

Sarin

ζ Her

NE

LYRA

BOÖTES

M13

Vega

CYGNUS

Izar

VULPECULA

Arcturus

Deneb

Elephant's Trunk Nebula

Eltanin

Iris Nebula

VIRGO

DRACO

Alderamin

Vindemiatrix

CEPHEUS

CASSIOPEIA

CANES VENATICI

Zenith

Alkaid

Caph

Mizar

LACERTA

Alioth

COMA BERENICES

Kochab

Shedar

N

Dubhe

URSA MINOR

CAMELOPARDALIS

LEO MINOR

Polaris

Merak

URSA MAJOR

PERSEUS

Mirfak

LEO

Algieba

LYNX

Algol

SEXTANS

α Lyn

Regulus

Capella

Mars

Castor

GEMINI

AURIGA

CANCER

Pollux

Elnath

Beehive Cluster

Jupiter

Alhena

Ecliptic

NW

CANIS MINOR

Procyon

ORION

W

MONOCEROS

HYDRA

CONSTELLATIONS

Galaxies ○
Star Clusters ○
Nebulae ○
Planets ● ● ● ●
Stars ● ● ·
Region of interest

May 1: 11:00 p.m. (12:00 a.m. DST)
May 15: 10:00 p.m. (11:00 p.m. DST)
June 1: 9:00 p.m. (10:00 p.m. DST)

Iris Nebula (NGC 7023)

Highlights in the Northern Sky

Cassiopeia (the Queen) is now hugging the northern horizon, with the house-like **Cepheus (the King)**, now appearing to turn sideways. You can find the **Iris Nebula (NGC 7023)** just above Cepheus's box. This bright reflection nebula (meaning its gases are illuminated by a star) lies about 1,300 light-years away and was discovered by German-British astronomer William Herschel in 1794.

Auriga (the Charioteer) is now low in the northwest with its prominent star **Capella** dominating the sky in that area. Though Capella appears as a single star, it is actually a quadruple star system that is roughly 43 light-years away.

Hercules is now well placed in the northeast, standing out with its keystone-shaped body. Hercules is home to one of the most magnificent globular star clusters in the northern sky, **Messier 13**. This stunning cluster has a magnitude of 5.8 and is roughly 25,000 light-years away. Believed to be about 12 billion years old and containing some 300,000 stars, it can be found easily with binoculars and is a stunning sight in even small telescopes.

Also beginning to rise in the northeast are the notable constellations **Cygnus (the Swan)** and **Lyra (the Harp)**. Lyra's brightest star is **Vega**, which by now is easily visible in the northeast. The star was made famous in Carl Sagan's 1985 book *Contact*; it was where an alien signal was picked up by telescopes on Earth.

June

June's Events

Summer begins this month, heralding in warmer nights and clearer skies.

While the planets have ruled the night sky since the beginning of the year, it's the morning sky where you can now enjoy a few planetary treats. Venus can be found low in the eastern sky before sunrise, with Saturn and Neptune higher in the southeast. While Venus and Saturn can be seen with the naked eye, it's best to use large binoculars to spot Neptune.

While Jupiter has sunk below the horizon after sunset, Mars is still visible as it cruises through Leo. On June 15, the Red Planet will lie less than half a degree from Regulus, Leo's brightest star. By the end of June, Mars will be very low on the western horizon.

This is also a great month to try to see noctilucent clouds, clouds that form high in the atmosphere and occur in the Northern Hemisphere beginning around the middle of May. They are best seen while the sun is below the horizon just after sunset or before sunrise. The nature of these clouds isn't completely understood. However, it is believed that they form from the dust of meteoroids on water-ice crystals high enough up in our atmosphere to catch the Sun's rays. While they used to only be seen closer to the polar regions, in recent years they have been consistently spotted farther south.

The summer solstice falls on June 20.

Calendar of Events

Day	Time (UTC)	Event
01	9:49	Mars 1.5°S of the Moon
03	3:41	First quarter of the Moon
07	6:42	Moon at apogee: 405,600 km (252,028 mi.)
10	10:25	Antares 0.3°N of the Moon
11	7:44	Full Moon
18	19:19	Last quarter of the Moon
19	3:47	Saturn 3.4°S of the Moon
20	22:42	Summer solstice
23	0:43	Moon at perigee: 363,200 km (225,682 mi.)
23	2:59	Pleiades 0.6°S of the Moon
25	10:31	New Moon
27	6:02	Mercury 2.9°S of the Moon
27	17:42	Beehive 2.1°S of the Moon
30	1:05	Mars 0.2°S of the Moon

The Moon This Month

SUN	MON	TUES	WED	THURS	FRI	SAT
1	2	3 1st Quarter	4	5	6	7
8	9	10	11 Full Moon	12	13	14
15	16	17	18 Last Quarter	19	20	21
22	23	24	25 New Moon	26	27	28
29	30					

The Moon swings by a few notable stars this month. On June 1, the waxing crescent Moon will be found near Regulus, the brightest star in Leo. Then, on June 5 and 6, the waxing gibbous Moon will hop past Spica, Virgo's brightest star. And on June 9, the Moon will shine near Antares, the brightest star in Scorpius, though the pair will be low in the south.

The Moon also visits some planets this month. On June 18 to 19, the waning gibbous Moon will join Saturn and Neptune in the early morning sky. Finally, on June 29, there will be a very close conjunction of Mars and the waning crescent Moon in the western sky after sunset.

The Moon will be at first quarter on June 3, and the full Moon falls on June 11. Last quarter is on June 18, while the new Moon falls on June 25. Apogee is on June 7, and the Moon will be at perigee on June 23.

The constellation Leo. The Moon will be near its brightest star, Regulus, on June 1

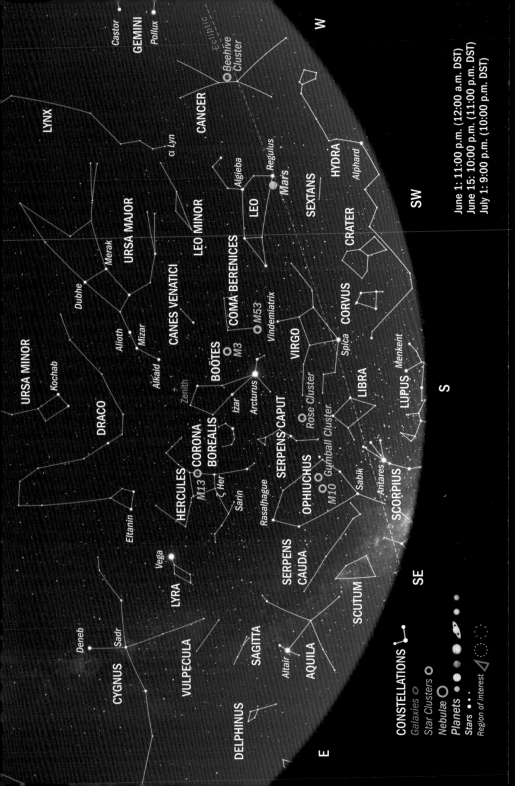

Messier 13

Highlights in the Southern Sky

Looking to the south, **Boötes (the Herdsman)** is high in the sky, with its star **Arcturus** shining brightly.

Next to Boötes is **Canes Venatici (the Hunting Dogs)**, where you can find **Messier 3**, a stunning globular cluster. South of it, **Coma Berenices (Berenice's Hair)** contains another wonderful globular cluster, **Messier 53**.

Virgo (the Maiden) sits south of Coma Berenices. Virgo's brightest star, **Spica**, is a blue-white star that lies 250 light-years away. Like many stars, Spica is a binary star, though its companion cannot be distinguished visually, even with the largest telescopes.

For more globular clusters, if you turn toward the southeast, you'll see the well-placed **Rose Cluster (Messier 5)** in **Serpens Caput (the Head of the Serpent)**. For all of these clusters, binoculars will reveal their fuzzy components, but even a small telescope will reveal their stars more clearly.

Ophiuchus (the Serpent Bearer) can be found in the southeast with its brightest star, **Rasalhague**. It, too, has a couple of notable globular clusters: **Messier 10** and the brighter **Gumball Cluster (Messier 12)**. This cluster lies roughly 15,500 light-years away and is best seen through medium-to-large telescopes if you'd like to distinguish between individual stars.

Hercules is high in the southeast and is where you can find **Messier 13**, another impressive globular cluster.

Finally, the two most prominent summer constellations — **Sagittarius (the Archer)** and **Scorpius (the Scorpion)** — are beginning to rise over the southern horizon with their treasure trove of objects. The pair will become more prominent in the following months.

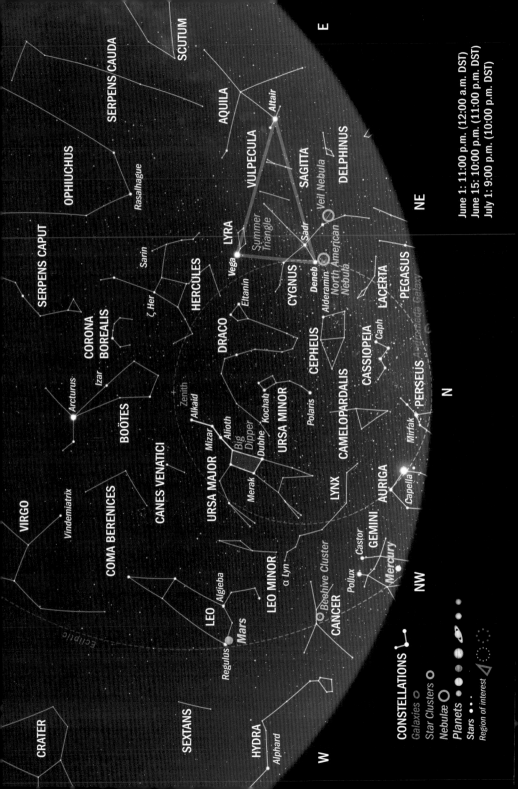

Eastern Veil Nebula (NGC 6992)

Highlights in the Northern Sky

You may have noticed a bright star in the northeast. That would be **Deneb**, which is the brightest star in **Cygnus (the Swan)**. The star is a blue-white supergiant that lies roughly 2,000 light-years away. It has a mass 19 times that of our own Sun. It is most notable as part of an asterism called the **Summer Triangle**, which includes two other stars: **Vega** and **Altair**.

Vega can be found in **Lyra (the Harp)** shining brightly near Cygnus, while Altair lies in **Aquila (the Eagle)** in the east.

Cygnus is a wonderful area for those with binoculars, as it is thick with stars, including many open clusters. The constellation is also home to the **North America Nebula (NGC 7000)**, which is best seen using medium-to-large telescopes. Cygnus also contains the **Veil Nebula**, which is broken up into the **Eastern Veil Nebula (NGC 6992)** and the **Western Veil Nebula (NGC 6960)**. The Veil Nebula is a massive supernova remnant that stretches 110 light-years across, taking up six Moon widths of the sky.

Ursa Minor (the Little Bear) now looks like an upside-down spoon in the north. **Ursa Major (the Great Bear)** is found in the northwest, and the handle of its asterism the **Big Dipper** rises high in the sky.

July

July's Events

We are now into the summer months — when the kids are out of school and many people are thinking about taking vacations. If you can get out of the city and away from light pollution, July is a perfect opportunity to enjoy the night sky, especially as the nights are getting marginally longer.

The richness of the summer constellations is upon us. The center of our galaxy can be found in the south, where Sagittarius and Scorpius dominate.

Mars is now sinking low in the west, which means if you want to see some planets, you'll have to either go to bed in the wee hours of the morning or get up just before sunrise. But if you are up for that, you can find Venus shining brightly in the morning sky, along with Saturn. Faint Neptune can also be found close to Saturn, though you'll need a telescope to see it best.

Earth is at aphelion, or the farthest from the Sun in its annual orbit, on July 3.

Calendar of Events

Day	Time (UTC)	Event
02	19:31	First quarter of the Moon
03	17:59	Earth at aphelion: 1.0166 astronomical units (au)
04	3:59	Mercury 25.9°E of the Sun (greatest eastern elongation)
04	22:28	Moon at apogee: 404,600 km (251,407 mi.)
10	20:37	Full Moon
13	4:29	Aldebaran 3.1°N of Venus
16	10:19	Saturn 3.9°S of the Moon
18	0:38	Last quarter of the Moon
20	9:27	Pleiades 0.8°S of the Moon
20	9:52	Moon at perigee: 368,000 km (228,665 mi.)
21	21:00	Venus 6.5°S of the Moon
23	4:20	Jupiter 5°S of the Moon
24	19:11	New Moon
28	19:45	Mars 1.4°N of the Moon
31	5:45	Spica 1.1°N of the Moon

The Moon This Month

SUN	MON	TUES	WED	THURS	FRI	SAT
		1	2 1st Quarter	3	4	5
6	7	8	9	10 Full Moon	11	12
13	14	15	16	17	18 Last Quarter	19
20	21	22	23	24 New Moon	25	26
27	28	29	30	31		

The Moon will shine near Spica, the brightest star in Virgo, on July 3, and then it will be near Antares, the brightest star in Scorpius, on July 7.

On July 16, there will be a conjunction of the waning gibbous Moon and Saturn in the early-morning hours. Neptune will be between the Moon and the ringed planet, but it will only be visible through telescopes.

The waning crescent Moon will join Venus and Jupiter in the morning sky on July 21 and 22. The trio will be best seen an hour or so before sunrise.

A week later, on July 28, there will be an evening conjunction of Mars and the thin waxing crescent Moon.

First quarter falls on July 2, and the full Moon is on July 10. It's the last quarter on July 18, and the new Moon rises on July 24. The Moon will be at apogee on July 4 and at perigee on July 20.

Crescent Moon

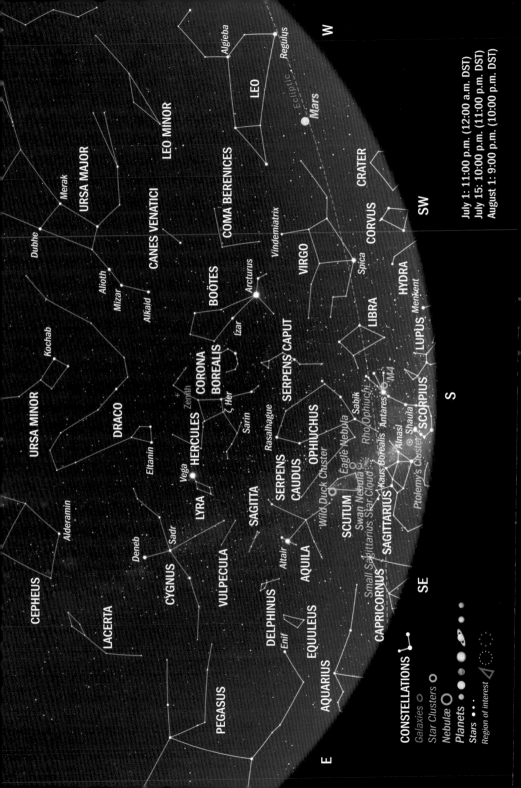

W

Algieba

Regulus

LEO

Ecliptic

Mars

Merak

URSA MAJOR

LEO MINOR

Dubhe

CANES VENATICI

COMA BERENICES

CRATER

Alioth

Mizar

Vindemiatrix

CORVUS

Alkaid

BOÖTES

Arcturus

VIRGO

Spica

SW

Izar

Kochab

CORONA
BOREALIS

ζ Her

Zenith

Sarin

SERPENS CAPUT

HYDRA

LIBRA

LUPUS   Menkent

URSA MINOR

DRACO

Rasalhague

OPHIUCHUS

M4

SCORPIUS

July 1: 11:00 p.m. (12:00 a.m. DST)
July 15: 10:00 p.m. (11:00 p.m. DST)
August 1: 9:00 p.m. (10:00 p.m. DST)

Eltanin

HERCULES

Sabik

Antares

Rho Ophiuchi

Shaula

Alderamin

Vega

LYRA

SERPENS
CAUDA

Eagle Nebula

Alnasl

Kaus Borealis

Ptolemy's Cluster

S

Deneb

Sadr

SAGITTA

Wild Duck Cluster

SCUTUM

Swan Nebula

Small Sagittarius Star Cloud

CEPHEUS

CYGNUS

VULPECULA

Altair

AQUILA

SAGITTARIUS

LACERTA

DELPHINUS

Enif

EQUULEUS

CAPRICORNUS

SE

PEGASUS

AQUARIUS

E

CONSTELLATIONS

Galaxies

Star Clusters

Nebulæ

Planets

Stars

Region of interest

Highlights in the Southern Sky

The richest part of the night sky now lies in the south, with the thickest part of our Milky Way galaxy dominating the view. In fact, if you'd like to gaze toward the center of our galaxy, this is the perfect time to do so.

It's difficult to know where to even start when pointing out the many wonderful targets in this region. The best part is you don't necessarily need a telescope to enjoy them. A decent pair of binoculars will do the trick and likely leave you in awe.

Let's start with **Scorpius (the Scorpion)**. It can be found in the south, with its brightest star, **Antares**, standing out nicely. The star is an aging red supergiant, much like Orion's Betelgeuse. Around Antares is a stunning region of gas known as the **Rho Ophiuchi cloud complex**. It lies just 460 light-years away and is the closest star-forming region to our Sun. Its rich multitude of colors makes it a particularly popular astro-photography target.

Just to the right of Antares is **Messier 4**, a beautiful globular cluster. There are more open clusters in the surrounding area that are easily seen with binoculars.

Also found in the south is **Sagittarius (the Archer)**. It can be made out by its teapot shape. This area is teeming with open and globular clusters. At the tip of the teapot's spout is the star **Alnasl**. The spout pours in the direction of the center of our galaxy. Nearby, in between Alnasl and **Shaula**, which marks the tail of Scorpius, is **Ptolemy's Cluster (Messier 7)**, a stunning open cluster.

At the top of Sagittarius's teapot is the star **Kaus Borealis**, the constellation's brightest star. If you scan upward from there using binoculars, you will see the prominent **Small Sagittarius Star Cloud (Messier 24)**. A bit above that is the **Swan Nebula (Messier 17)**, which also goes

Rho Ophiuchi cloud complex

by the name the **Omega Nebula**. And further north still, you can find the **Eagle Nebula (Messier 16)**. This gas cloud contains the "Pillars of Creation," a name given to it after the Hubble Space Telescope imaged it in 1995. The James Webb Space Telescope captured another view of it in 2022.

To the upper left of Sagittarius is **Scutum (the Shield)**, where you can find the **Wild Duck Cluster (Messier 11)**, an open cluster named for the roughly V-shaped arrangement of its brightest stars.

Whirlpool Galaxy (Messier 51)

Highlights in the Northern Sky

In the northwest sits **Ursa Major (the Great Bear)**. Try to test your eyesight and see if you can see both **Alcor** and **Mizar** — a pair of stars in the **Big Dipper's** handle — with your naked eye. If you can't separate the pair unaided, you can use modest binoculars to do so. There are many galaxies to be found in this region, but most are best viewed with medium-to-large telescopes. Near the bear's tail is **Alkaid**, and there are two beautiful galaxies found nearby: the **Pinwheel Galaxy (Messier 101)** and the **Whirlpool Galaxy (Messier 51)**.

Hercules, where there are two fantastic globular clusters, **Messier 13** and **Messier 92**, sits near the zenith.

Turning to the northeast, **Cepheus (the King)** lies high in the sky in late evening. A good pair of binoculars will lead you to the small open **Swimming Alligator Cluster (NGC 7160)** located just inside Cepheus's house shape. And Cepheus's **Garnet Star** nearby is worth a look. Also nearby is the **Elephant's Trunk Nebula (IC 1396)** as well as the **Iris Nebula (NGC 7023)** and the **Wizard Nebula (NGC 7380)**.

Lacerta (the Lizard) to the right of Cepheus looks like a tiny kite in the sky. South of Cepheus is **Cassiopeia (the Queen)**. This constellation has many open clusters, which are a binocular treat. It is also home to the **Heart Nebula (IC 1805)** and the **Soul Nebula (IC 1848)**, two faint nebulae that are often photographed together.

The constellation **Andromeda (the Princess)** located south of Cassiopeia stretches into **Pegasus (the Flying Horse)**, which is in the east. There's a chance you may have noticed a fuzzy patch in Andromeda, low in the northeastern sky. That would be the **Andromeda Galaxy (Messier 31)**, which is visible with the naked eye from dark-sky locations. If you're trying to look for it, you can search above **Mirach**, Andromeda's second-brightest star. The upper three stars of Cassiopeia form an arrow pointing at Messier 31. However, the galaxy will be somewhat lower in the sky until the end of the month.

August

August's Events

Venus and Jupiter shine in the morning sky, joining Saturn and faint Neptune, which rise around midnight. Mercury can also be found low in the east around the middle of August. On August 12, there is a spectacular conjunction between Jupiter and Venus in the eastern morning sky before sunrise. The pair will be separated by less than a quarter of a degree.

Meanwhile, a wealth of objects is still visible in the night sky, with the center of the Milky Way taking front stage. Check out the southern sky to see an abundance of star clusters and nebulae.

Finally, the Perseid meteor shower occurs this month. This is considered the best meteor shower of the year for those in the Northern Hemisphere. This year, the Perseids peak on the night of August 12 to 13. However, you will have to battle the waning gibbous Moon.

Calendar of Events

Day	Time (UTC)	Event
01	12:41	First quarter of the Moon
01	16:37	Moon at apogee: 404,200 km (251,158 mi.)
04	1:40	Antares 0.6°N of the Moon
09	7:55	Full Moon
12	6:53	Jupiter 0.9°N of Venus
12	15:05	Saturn 4.1°S of the Moon
12		Perseid meteor shower peak
14	14:01	Moon at perigee: 369,300 km (229,472 mi.)
16	5:12	Last quarter of the Moon
16	16:09	Pleiades 0.9°S of the Moon
19	9:59	Mercury 18.6°W of the Sun (greatest western elongation)
19	21:05	Jupiter 4.9°S of the Moon
20	10:51	Venus 5°S of the Moon
20	12:07	Pollux 2.5°N of the Moon
21	11:09	Beehive 2.1°S of the Moon
23	6:06	New Moon
26	15:41	Mars 3°N of the Moon
29	11:34	Moon at apogee: 404,600 km (251,407 mi.)
31	6:25	First quarter of the Moon
31	9:55	Antares 0.7°N of the Moon

The Moon This Month

SUN	MON	TUES	WED	THURS	FRI	SAT
					1 1st Quarter	2
3	4	5	6	7	8	9 Full Moon
10	11	12	13	14	15	16 Last Quarter
17	18	19	20	21	22	23 New Moon
24	25	26	27	28	29	30
31 1st Quarter						

On August 12 a conjunction between Saturn and the waning gibbous Moon will be visible all night long. Neptune is also part of the conjunction; however, the icy blue giant is only visible through a telescope.

On August 19, there will be a beautiful sight in the eastern morning sky before sunrise: The waning crescent Moon will line up nicely with Jupiter and Venus. The next morning, the Moon, Venus and Jupiter will form a nice triangle.

On August 21, quite low in the eastern morning sky, there is a conjunction with the waning crescent Moon above Mercury.

Meanwhile, the Moon and Mars meet up in the western sky on August 26 after sunset. However, you'll need a very good view of the western horizon as the pair will be quite low.

First quarter falls on August 1 and August 31, while the full Moon occurs on August 9. Last quarter is on August 16. The new Moon is on August 23. The Moon will be at apogee twice this month: August 1 and 29. It will be at perigee on August 14.

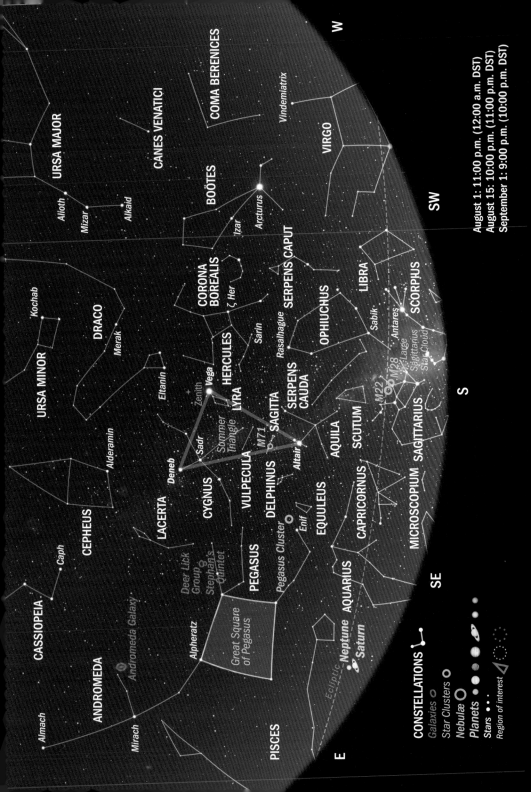

Deer Lick Group (NGC 7331 Group) and the nearby Stephan's Quintet (on the right)

Highlights in the Southern Sky

Pegasus (the Flying Horse) is rising in the east. As one of the largest constellations in the sky, it's easy to identify by its large square asterism, referred to simply as the **Great Square of Pegasus.** The head of Pegasus is the star **Enif,** which you can use to point you to an impressive globular cluster, **Messier 15,** known as the **Pegasus Cluster.**

Also within the constellation is a beautiful cluster of galaxies called **Stephan's Quintet.** Four of the galaxies are interacting with one another at a distance of about 300 million light-years, with the fifth being a visual companion that is much closer to Earth, about 39 million light-years away.

Another group of galaxies lie nearby called the **Deer Lick Group (NGC 7331 Group).** There are four galaxies in this small cluster, though one foreground galaxy, **NGC 7331,** is much larger and closer.

You can find three small constellations near Pegasus: **Sagitta (the Arrow), Delphinus (the Dolphin)** and **Equuleus (the Foal).** Also look for **Messier 71,** a nice globular cluster in Sagitta.

You may have noticed three very bright stars at the zenith. They form the asterism of the **Summer Triangle,** which is part of three separate constellations. **Altair** is one of the stars, found in **Aquila (the Eagle),** along with **Deneb** in **Cygnus (the Swan)** and **Vega** in **Lyra (the Harp).**

Swinging back to the south, **Sagittarius (the Archer)** is very well placed this month. Using binoculars, you can see the globular clusters **Messier 22 (the Great Sagittarius Cluster)** and **Messier 28** at the top of the constellation's teapot-shaped asterism. In the direction of the teapot's spout is the **Large Sagittarius Star Cloud (Messier 24)** and several open clusters.

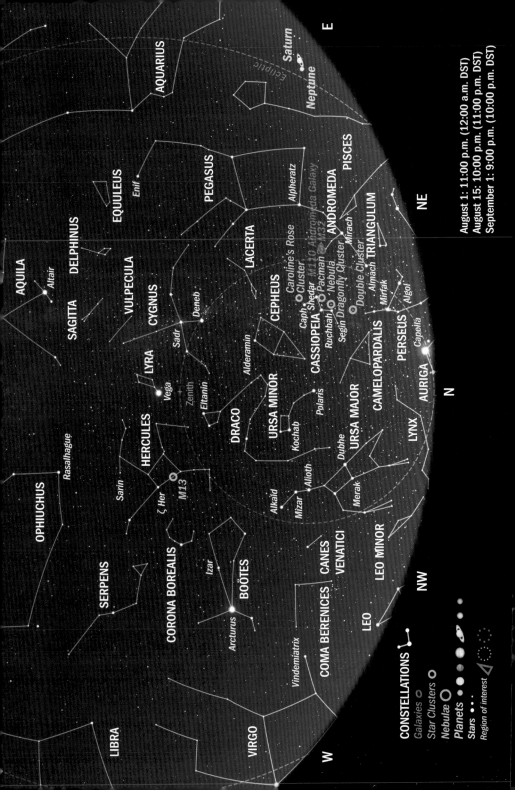

Double Cluster (NGC 869 and NGC 884)

Highlights in the Northern Sky

In the northeast, you will find the familiar W-shaped **Cassiopeia (the Queen)** with several open clusters, including **Caroline's Rose Cluster (NGC 778)**, which can be found near the second-brightest star in the constellation, **Caph**. Near the star **Schedar** lies the **Pacman Nebula (NGC 281)**. The impressive star-forming region is located 9,200 light-years away and is 48 light-years across.

There are other open clusters on view, including the **Dragonfly Cluster (NGC 457)**, which is visible through binoculars, along with several more visible between Cassiopeia's stars **Segin** and **Ruchbah**.

Perseus (the Hero) is beginning to rise higher in the northeast, which provides amateur astronomers with a perfect opportunity to see the **Double Cluster (NGC 869 and NGC 884)** above him. The pair of open clusters can be seen with the naked eye from dark-sky locations, but binoculars best reveal their beauty.

Andromeda (the Princess) is also visible in the northeast with the **Andromeda Galaxy (Messier 31)**, a massive barred spiral galaxy that will collide with our own Milky Way in billions of years. A small telescope will also reveal two of its neighboring galaxies, **Messier 32** and **Messier 110**.

Cepheus (the King) now looks like an upside-down house in the north. **Draco (the Dragon)** is winding to the upper left of **Ursa Minor (the Little Bear)** and **Polaris. Ursa Major (the Great Bear)** is now low in the northwest.

Turning to the west, you'll find **Boötes (the Herdsman)** with its star **Arcturus** shining brightly. The constellation is beginning to sink lower and will disappear in the coming months.

Finally, **Hercules** is high in the west, which means **Messier 13** — the best globular star cluster in the Northern Hemisphere — is in a prime viewing location for those with binoculars or small telescopes.

September

September's Events

We are beginning the gradual transition from the summer constellations to the winter ones. The Summer Triangle is still prominent, with the stars Deneb, Vega and Altair shining high above. Meanwhile, Sagittarius and Scorpius are beginning to sink in the southwest, though Sagittarius can still be viewed in the south after dusk.

Saturn is beginning to rise in the east after sunset, with Neptune nearby. Jupiter also rises in the east well after midnight. Uranus rises in the northeast in the late evening.

The autumn equinox falls on September 22.

Calendar of Events

Day	Time (UTC)	Event
07	18:09	Full Moon
07	18:11	Total lunar eclipse (not observable from North America)
08	20:09	Saturn 4°S of the Moon
10	8:10	Moon at perigee: 364,800 km (226,676 mi.)
12	21:48	Pleiades 1°S of the Moon
12	14:28	Spica 2.2°S of Mars
14	10:33	Last quarter of the Moon
16	11:06	Jupiter 4.7°S of the Moon
16	17:58	Pollux 2.5°N of the Moon
17	19:24	Beehive 2.1°S of the Moon
19	9:59	Regulus 0.5°N of Venus
19	11:11	Regulus 1.4°S of the Moon
19	11:46	Venus 0.9°S of the Moon
21	5:10	Saturn at opposition
21	19:42	Partial solar eclipse (not observable from North America)
21	19:54	New Moon
22	14:20	Autumn equinox
23	11:29	Neptune at opposition
23	23:31	Spica 1.2°N of the Moon
24	12:50	Mars 4.3°N of the Moon
26	5:46	Moon at apogee: 405,600 km (252,028 mi.)
27	17:34	Antares 0.7°N of the Moon
29	23:54	First quarter of the Moon

The Moon This Month

SUN	MON	TUES	WED	THURS	FRI	SAT
	1	2	3	4	5	6
7 Full Moon	8	9	10	11	12	13
14 Last Quarter	15	16	17	18	19	20
21 New Moon	22	23	24	25	26	27
28	29 1st Quarter	30				

On September 8, the almost-full Moon will join Saturn in the sky all night long.

On the night of September 12 to 13, you can find the waning gibbous Moon near the Pleiades. A few days later, in the early morning hours of September 16, you can find Jupiter and the waning crescent Moon in the east.

Finally, on September 19, there will be a gorgeous close conjunction between the thin crescent Moon and Venus. In parts of Canada's far north, the Moon will even occult (pass in front of) the planet. You will also find Regulus, the brightest star in Leo, near them.

The full Moon rises on September 7, with last quarter on September 14. The new Moon occurs on September 21, and first quarter falls on September 29. The Moon is at perigee on September 10 and at apogee on September 26.

Stunning "moonbows" like this occur when the Moon's light is refracted through water droplets in the air

W

BOÖTES

Alkaid

Arcturus

Izar

CORONA BOREALIS

SERPENS CAPUT

SW

Sarin

○ M13

ζ Her

HERCULES

Rasalhague

OPHIUCHUS

○ M10

Scutum Star Cloud
Wild Duck Cluster

Eagle Nebula
Swan Nebula

Polis   *Sabik*
Trifid Nebula
Lagoon Nebula

Eltanin

DRACO

URSA MINOR

Vega

LYRA

VULPECULA

SAGITTA

SERPENS
CAUDA

β Scuti

AQUILA

SCUTUM

Snail
Sagittarius Star
Cloud ☉ M22

SAGITTARIUS

Kaus Borealis

S

Alderamin

Sadr

Zenith

Deneb

CYGNUS

Summer Triangle

Altair

Enif

DELPHINUS

Pegasus Cluster ○

EQUULEUS

CAPRICORNUS

MICROSCOPIUM

CEPHEUS

Caph

LACERTA

Alpheratz

PEGASUS

Markab

AQUARIUS

PISCIS AUSTRINUS

Fomalhaut

CASSIOPEIA

Almach

ANDROMEDA

Andromeda Galaxy

Mirach

PISCES

Neptune

Saturn

SE

Diphda

Mirfak

PERSEUS

Algol

TRIANGULUM

ARIES

Hamal

Ecliptic

CETUS

E

September 1: 11:00 p.m. (12:00 a.m. DST)
September 15: 10:00 p.m. (11:00 p.m. DST)
October 1: 9:00 p.m. (10:00 p.m. DST)

CONSTELLATIONS ⌐
Galaxies ○
Star Clusters ○
Nebulae ◯
Planets ● ● ●
Stars ● • ·
Region of interest △

Highlights in the Southern Sky

In the southwestern sky, **Sagittarius (the Archer)** is beginning to sink lower on the horizon. You can still get good views of the **Small Sagittarius Star Cloud (Messier 24)**, found near the star **Polis** with the **Swan Nebula (Messier 17)** just north of the star cloud.

The great Sagittarius globular cluster, **Messier 22**, is north of the tip of the constellation's teapot asterism, above the star **Kaus Borealis**. Even modest binoculars will provide an excellent view. The **Lagoon Nebula (Messier 8)** and the **Trifid Nebula (Messier 20)** are to the right of that star.

Scutum (the Shield) can be found north of Sagittarius with its open cluster the **Wild Duck Cluster (Messier 11)** near the constellation's second-brightest star, **Beta Scuti**. There is also the **Scutum Star Cloud** found in the same location. All of these are excellent binocular targets. The **Eagle Nebula (Messier 16)** can be located to the right of the base of Scutum, but in the constellation **Serpens Cauda (the Tail of the Serpent)**.

Ophiuchus (the Serpent Bearer) separates the reptilian constellation Serpens into two parts: **Serpens Cauda (the Tail of the Serpent)** and **Serpens Caput (the Head of the Serpent)**. You can find the **Gumball Cluster (Messier 12)** in the center of Ophiuchus along with another great globular cluster, **Messier 10**.

Hercules is very well placed high in the southwest, with the impressive **Messier 13**, sometimes referred to as the **Hercules Cluster**, in an excellent location for those with binoculars and small telescopes.

Also high in the south is the **Summer Triangle** asterism, which is made up of three stars from three separate constellations: **Deneb**, the brightest star in **Cygnus (the Swan)**; **Vega**, the brightest star in **Lyra (the Harp)**;

Trifid Nebula (Messier 20)

and **Altair**, the brightest star in **Aquila (the Eagle)**.

Aquarius (the Water Bearer) is low in the southeastern sky to the lower right of **Pegasus (the Flying Horse)**. The **Pegasus Cluster (Messier 15)** is found above the star **Enif** and very close to the small constellation **Equuleus (the Foal)**.

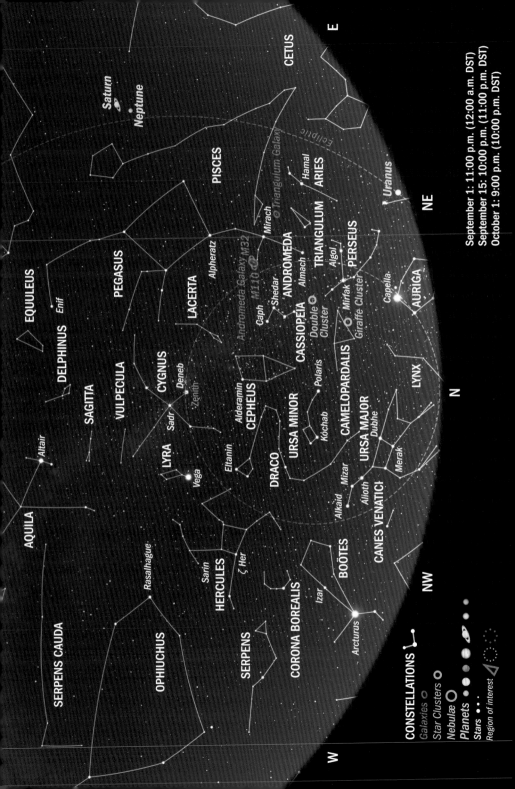

Triangulum Galaxy (Messier 33)

Highlights in the Northern Sky

In the northwest, **Boötes (the Herdsman)** is beginning to sink lower, with its shining bright star **Arcturus** straddling the west. **Ursa Major (the Great Bear)** is also quite low on the northwestern horizon. **Draco (the Dragon)** winds around **Ursa Minor (the Little Bear)**.

House-shaped **Cepheus (the King)** is now upside-down. **Cassiopeia (the Queen)** is a sideways "W" or "M" in the northeast. Its brightest star, **Schedar**, points toward the **Andromeda Galaxy (Messier 31)** and its satellite galaxies, **Messier 32** and **Messier 110**.

South of the constellation **Andromeda (the Princess)** is the tiny constellation **Triangulum (the Triangle)**. This is where you can find a favorite galaxy of many amateur astronomers called the **Triangulum Galaxy (Messier 33)**. Though it is considered a bright galaxy, it is best seen through binoculars and low-power telescopes due to its low surface brightness and large diameter.

South of Cassiopeia is **Perseus (the Hero)**. Almost halfway between the constellations you can find the two open clusters known collectively as the **Double Cluster (NGC 869 and NGC 884)**.

Camelopardalis (the Giraffe) lies to the west of Perseus. Here, you can find the open cluster known as the **Giraffe Cluster (NGC 1502)**. Also called the **Jolly Roger Cluster**, this is a young cluster of about 60 stars that is about 3,500 light-years from Earth.

By the middle of the month, if you're up late, you may notice a very bright star beginning to rise in the northwest. That would be **Capella,** the brightest star in **Auriga (the Charioteer)**. The constellation will rise higher in the northeast by the end of the month.

October

October's Events

This month several bright planets are once again dominating the night sky.

Saturn, which reached opposition in September, is now shining brightly in Aquarius, while Jupiter rises in the east after midnight. Neptune can also be found near our Solar System's ringed planet, though only through telescopes (it can be seen through even small ones). Uranus is in Taurus, not too far from the Pleiades. A good pair of binoculars or a small telescope will show the icy giant. Mars is still lost in the evening Sun's glare. Mercury will join it at the end of the month. Venus is still gracing the morning sky before sunrise.

Fall rings in several year-end meteor showers, beginning this month with the Orionids.

It's not a spectacular shower, though under this year's moonless sky conditions it can produce about 10 to 20 meteors an hour on its peak night, which will be October 21 to 22.

Calendar of Events

Day	Time (UTC)	Event
05	22:46	Saturn 3.8°S of the Moon
07	3:48	Full Moon
08	8:36	Moon at perigee: 359,800 km (223,569 mi.)
10	4:20	Pleiades 0.9°S of the Moon
13	18:13	Last quarter of the Moon
13	22:31	Jupiter 4.4°S of the Moon
13	23:30	Pollux 2.6°N of the Moon
14	22:53	Beehive 2°S of the Moon
19	21:37	Venus 4°N of the Moon
21		Orionid meteor shower peak
21	12:25	New Moon
23	12:00	Mars 4.8°N of the Moon
23	15:15	Mercury 2.5°N of the Moon
23	19:31	Moon at apogee: 406,400 km (252,525 mi.)
25	2:15	Antares 0.6°N of the Moon
29	16:21	First quarter of the Moon
29	16:59	Mercury 23.9°E of the Sun (greatest elongation east)

The Moon This Month

SUN	MON	TUES	WED	THURS	FRI	SAT
			1	2	3	4
5	6	7 **Full Moon**	8	9	10	11
12	13 **Last Quarter**	14	15	16	17	18
19	20	21 **New Moon**	22	23	24	25
26	27	28	29 **1st Quarter**	30	31	

On October 5, the full Moon will be only 2 degrees from Saturn. On October 9, the waning gibbous Moon will cross the Pleiades. In the hours of October 14, the Moon will join Jupiter in the eastern sky. Pollux, the second-brightest star in Gemini, will also be nearby.

We get a full Moon on October 7 and the last quarter on October 13. The new Moon falls on October 21 and first quarter on October 29. The Moon is at perigee on October 8 and at apogee on October 23.

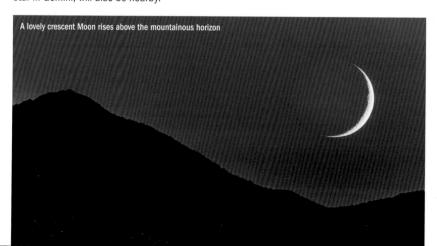

A lovely crescent Moon rises above the mountainous horizon

W

CORONA
BOREALIS

ζ Her

HERCULES

Sarin

Eltanin

Rasalhague

OPHIUCHUS

DRACO

Vega

Double Double
(ε Lyræ)

LYRA

Ring Nebula

SERPENS
CAUDA

SW

CEPHEUS

Alderamin

Sadr

VULPECULA

SAGITTA

SCUTUM

Deneb

CYGNUS

DELPHINUS

Altair

EQUULEUS

AQUILA

Caph

Zenith

LACERTA

Enif

M2

Sadalsuud

SAGITTARIUS

CASSIOPEIA

ANDROMEDA

Andromeda Galaxy

Alpheratz

Great
Square of
Pegasus

PEGASUS

Pegasus Cluster

Markab

CAPRICORNUS

Jellyfish
Cluster

S

Mirach

Almach

PISCES

Neptune

Saturn

AQUARIUS

PISCIS AUSTRINUS

Fomalhaut

TRIANGULUM

Hamal

ARIES

Diphda

SCULPTOR

CAMELOPARDALIS

Pleiades

Uranus

CETUS

Menkar

SE

Capella

PERSEUS

TAURUS

Hyades

Aldebaran

Elnath

ERIDANUS

AURIGA

Ecliptic

E

October 1: 11:00 p.m. (12:00 a.m. DST)
October 15: 10:00 p.m. (11:00 p.m. DST)
November 1: 9:00 p.m. (10:00 p.m. DST)

CONSTELLATIONS
Galaxies ○
Star Clusters ○
Nebulæ ○
Planets ● ● ●
Stars ● ● ·
Region of interest △

Ring Nebula (Messier 57)

Highlights in the Southern Sky

Saturn is well placed in the southern sky this month, lying just on the border of **Aquarius (the Water Bearer)**, even though the planet doesn't get particularly high.

Looking toward Aquarius, you can find **Messier 2**, a very nice globular cluster found to the right of the constellation's second-brightest star, **Sadalsuud**. The cluster is one of the largest of its kind, spanning roughly 175 light-years in diameter, and lies at a distance of 37,500 light-years. It was also the first globular cluster to be added to the Messier Catalog.

Pegasus (the Flying Horse) is unmistakable with its **Great Square of Pegasus** dominating the southern sky. Be sure to grab binoculars to take in the globular **Pegasus Cluster (Messier 15)**, which is in a prime viewing location for the entire month.

Pisces (the Fishes) winds its way between Pegasus and **Cetus (the Whale)**.

Capricornus (the Sea Goat) is a faint constellation that lies low in the southern sky to the lower right of Aquarius. Use large binoculars to find the **Jellyfish Cluster (Messier 30)** in the constellation. It's a bright globular cluster that lies roughly 27,000 light-years away.

Altair is shining in the southwest in **Aquila (the Eagle)**, though it's most certainly **Vega**, the brightest star in **Lyra (the Harp)**, that you may have noticed first. In Lyra lies the beautiful **Ring Nebula (Messier 57)**, which is best seen through medium-sized telescopes. Be sure to also check out the **Double Double**, Lyra's multi-star gem near Vega.

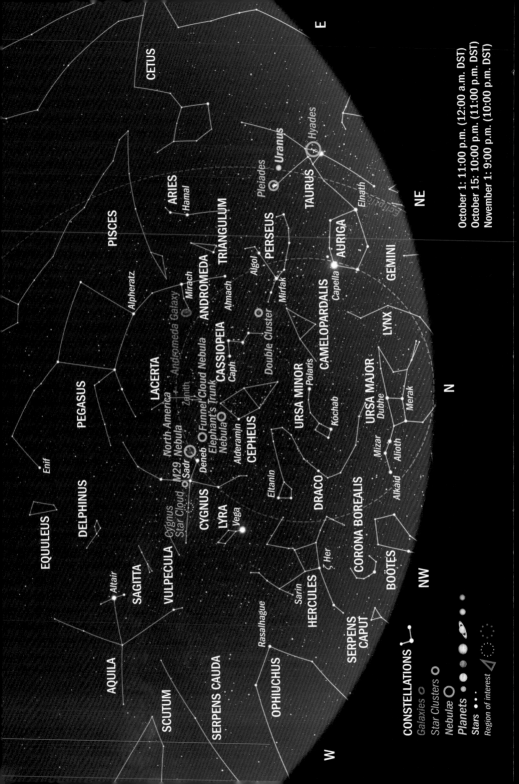

E

CETUS

ARIES
Hamal

Uranus
Hyades
Pleiades
TAURUS
Elnath
Ecliptic

PISCES

PERSEUS
Algol·
Mirfak
AURIGA
Capella

NE

ANDROMEDA
Almach
TRIANGULUM
Alpheratz
Mirach
Andromeda Galaxy
CAMELOPARDALIS
GEMINI

PEGASUS
LACERTA
Caph
CASSIOPEIA
Double Cluster
LYNX

Enif
Zenith
'Cloud Nebula
North America
Nebula
Polaris
URSA MINOR
Dubhe
Merak
URSA MAJOR

DELPHINUS
M29.
Nebula
Sadr
Deneb○Funnel
Elephant's Trunk
Nebula○
Alderamin
CEPHEUS
Kochab
N

EQUULEUS
Cygnus
Star Cloud
CYGNUS
LYRA
Vega
Eltanin
DRACO
Mizar
Alioth
Alkaid

Altair
VULPECULA
SAGITTA

AQUILA

SCUTUM

OPHIUCHUS

SERPENS CAUDA

Rasalhague

Sarin
HERCULES
ζ Her
CORONA BOREALIS
BOÖTES
NW

SERPENS
CAPUT

W

CONSTELLATIONS
Galaxies ○
Star Clusters ○
Nebulæ ○
Planets ● ● ●
Stars · • ●
Region of interest △

October 1: 11:00 p.m. (12:00 a.m. DST)
October 15: 10:00 p.m. (11:00 p.m. DST)
November 1: 9:00 p.m. (10:00 p.m. DST)

North America Nebula (NGC 7000)

Highlights in the Northern Sky

High in the northern sky is **Cygnus (the Swan)**, where you can find thick regions of stars using just binoculars. There's also the **Cygnus Star Cloud**, which is an incredibly impressive sight in binoculars. The Cygnus region is also one of the richest-known star-forming areas in our galaxy.

Interestingly, **Cygnus X-1** is also located in this region and can be found halfway along the swan's long neck. It is a binary star system and a strong source of X-rays. This was the first X-ray source discovered that suggested the presence of a black hole.

The **North America Nebula (NGC 7000)** is also well placed, found just off Cygnus's tail star, **Deneb**. While it is a large nebula, it's best seen under very dark skies due to its low surface brightness. **Messier 29**, a nice open cluster, can also be seen nearby.

The **Funnel Cloud Nebula (Le Gentil 3)** also lies near Deneb. This is a dark nebula that is best seen through binoculars at dark-sky locations or with a small telescope. It resembles its name, cutting through a bright portion of stars in Cygnus. Nearby is the **Elephant's Trunk Nebula (IC 1396)**, found in **Cepheus (the King)**. Both nebulae are in prime viewing locations this month.

And speaking of prime viewing, there's the **Andromeda Galaxy (Messier 31)**, which is nice and high in the east this month, along with the **Double Cluster (NGC 869 and NGC 884)** in **Perseus (the Hero)**. Both are great targets for binoculars. Of course, the farther away you are from city lights, the better.

Finally, **Ursa Major (the Great Bear)** and its Big Dipper asterism will lurk above the northern horizon. By the end of the month, you will see it begin to rise in the north once again.

November

November's Events

The winter constellations are beginning to make their way into the southern sky, albeit very early in the morning hours.

High in the south, Pegasus rules the sky, while Saturn sits below it in Aquarius. In the east, Auriga and its brightest star, Capella, are starting to rise. Orion, too, is beginning to inch its way higher into the late evening sky.

You can find Jupiter rising in the east late in the night, with Uranus in Taurus. Venus will spend November sinking toward the eastern horizon before sunrise.

This month there are not one, not two, but *three* meteor showers to take in. The Southern Taurids peak on the full-Moon night of November 5 to 6, the Northern Taurids peak on the darker night of November 12 to 13, and finally the Leonids peak on the moonless night of November 17 to 18.

Calendar of Events

Day	Time (UTC)	Event
01	20:04	Spica 3.5°N of Venus
02	9:46	Saturn 3.7°S of the Moon
05		Southern Taurid meteor shower peak
05	13:19	Full Moon
05	17:29	Moon at perigee: 356,800 km (221,705 mi.)
06	14:26	Pleiades 0.8°N of the Moon
08	17:00	Antares 3.6°SE of Mercury
10	5:40	Pollux 2.7°N of the Moon
10	6:56	Jupiter 4°S of the Moon
11	4:27	Beehive 1.8°S of the Moon
12	5:28	Last quarter of the Moon
12		Northern Taurid meteor shower peak
12	23:51	Regulus 1.1°S of the Moon
12	17:41	Mars 1.2°N of Mercury
17	9:11	Spica 1.3°N of the Moon
17		Leonid meteor shower peak
19	21:48	Moon at apogee: 406,700 km (252,712 mi.)
20	6:47	New Moon
28	6:59	First quarter of the Moon
29	18:08	Saturn 3.8°S of the Moon

The Moon This Month

SUN	MON	TUES	WED	THURS	FRI	SAT
						1
2	3	4	5 Full Moon	6	7	8
9	10	11	12 Last Quarter	13	14	15
16	17	18	19	20 New Moon	21	22
23	24	25	26	27	28 1st Quarter	29
30						

On November 1, there is a conjunction between Saturn and the waxing gibbous Moon. The pair will only be separated by roughly 3 degrees. If you miss it, the pair will still be fairly close together the following night.

On November 5, the full Moon will shine close to the Pleiades, as it will the following night. On November 9, the waning gibbous Moon will lie roughly 2 degrees from Jupiter late all night.

Before sunrise on November 18, the old crescent Moon will sit to Venus's upper right.

Finally, on November 29, the waxing gibbous Moon once again meets with Saturn, and the pair will be roughly 3 degrees apart and still close on the following night.

The full Moon falls on November 5, with the last quarter on November 12. The new Moon occurs on November 20 and the first quarter on November 28. The Moon is at perigee on November 5, the same night as the full Moon, causing larger tides. The Moon is at apogee on November 19.

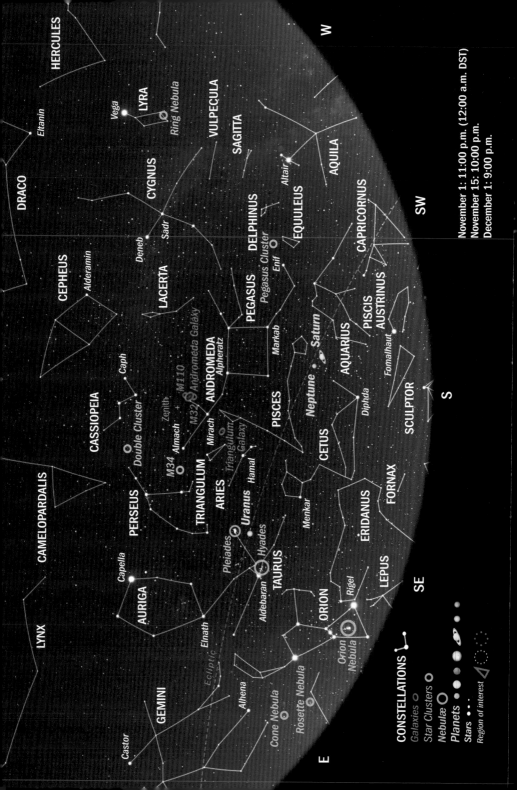

Pegasus Cluster (Messier 15)

Highlights in the Southern Sky

Pegasus (the Flying Horse) is still high in the southern sky this month. You can find the globular **Pegasus Cluster (Messier 15)** in the west of the constellation using binoculars.

Andromeda (the Princess) is also high in the southeast, where you can find the **Andromeda Galaxy (Messier 31)** with its two satellite galaxies, **Messier 32** and **Messier 110**. Nearby, in **Triangulum (the Triangle)**, sits the **Triangulum Galaxy (Messier 33)**, another impressive galaxy.

Perseus (the Hero) is climbing in the east, which means the **Double Cluster (NGC 869 and NGC 884)** is in a prime viewing location. **Messier 34**, also in Perseus, is a nice open cluster found 5 degrees above the star **Algol**.

When it comes to constellations that are commonly seen from the Southern Hemisphere, there are a few that can be spotted low in the southern sky, depending on your latitude. **Sculptor (the Sculptor's Workshop)** can be found above the southern horizon, along with **Piscis Austrinus (the Southern Fish)**. Piscis Austrinus is most known for its brightest star, **Fomalhaut**. Like so many stars in our sky, Fomalhaut isn't just a single star — it's actually a triple star system that lies roughly 25 light-years away.

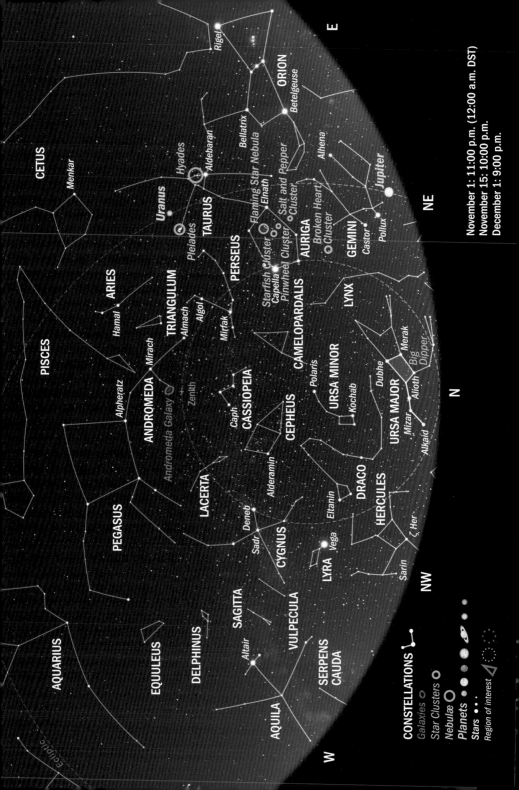

Flaming Star-Nebula (IC 405)

Highlights in the Northern Sky

Taurus (the Bull) is now in the northeast with two spectacular clusters. The **Hyades (Melotte 25)** is a stunning open cluster that is breathtaking in binoculars. Then there's the **Pleiades (Messier 45)**, another open cluster that is a favorite of many astrophotographers because of the blue hue of its dusty surroundings, which are lit up by the surrounding stars.

Gemini (the Twins) is creeping higher into the sky. You may also notice a particularly bright object not too far from Gemini's **Pollux**. It's not a star, but rather Jupiter, which has moved into the night sky.

Higher in the northeast is another bright object: **Capella**, the brightest star in **Auriga (the Charioteer)**. The constellation is home to many open clusters, including the **Starfish Cluster (Messier 38)**, the **Pinwheel Cluster (Messier 36)**, the **Salt and Pepper Cluster (Messier 37)** and the **Broken Heart Cluster (NGC 2281)**. There's also a nice nebula, known as the **Flaming Star Nebula (IC 405)**, that is both an emission and reflection nebula and lies roughly 1,500 light-years away.

Ursa Major (the Great Bear) is above the northeastern horizon, with its **Big Dipper** asterism looking like it's about to scoop up the sky.

In the northwest, try to take in **Cygnus (the Swan)** before it begins to drop too far in the west. It contains a multitude of open clusters that you can enjoy with binoculars. You can try to catch **Vega**, which shines brightly in **Lyra (the Harp)**, before it's gone. By the end of the month, it will be quite low in the west.

December

December's Events

December starts off with Mercury nicely visible in the east in the early morning sky just before sunrise. Saturn can still be found in the south once the Sun sets, with Neptune nearby. Jupiter rises later in the east, while Uranus can be found between them in Taurus with the aid of a small telescope.

One of the best meteor showers of the year, the Geminids can produce some 150 meteors an hour at their peak, which falls on the night of December 13 to 14. Though the Geminids occur when the weather is colder and cloudier, they rarely disappoint, producing impressive fireballs, which can be seen even from light-polluted cities.

There's also the Ursids, which peak on the night of December 22 to 23. This shower produces much fewer meteors — roughly 10 per hour on the peak night under this year's moonless, dark-sky conditions.

Calendar of Events

Day	Time (UTC)	Event
04	2:54	Pleiades 0.8°S of the Moon
04	6:06	Moon at perigee: 357,000 km (221,829 mi.)
04	23:14	Full Moon
07	16:48	Jupiter 3.7°S of the Moon
07	17:21	Pollux 2.9°N of the Moon
07	19:59	Mercury 20.7°W of the Sun (greatest elongation west)
08	16:23	Beehive 1.5°S of the Moon
10	5:32	Regulus 0.8°S of the Moon
11	20:52	Last quarter of the Moon
13		Geminid meteor shower peak
14	18:27	Spica 1.5°N of the Moon
17	1:09	Moon at apogee: 406,300 km (252,463 mi.)
18	11:00	Mercury 6.7°N of the Moon
19	9:07	Antares 5.5°N of Mercury
20	1:43	New Moon
21	15:03	Winter solstice
22		Ursid meteor shower peak
27	5:24	Saturn 4.2°S of the Moon
27	19:10	First quarter of the Moon
31	13:21	Pleiades 0.9°S of the Moon

The Moon This Month

SUN	MON	TUES	WED	THURS	FRI	SAT
	1	2	3	4 Full Moon	5	6
7	8	9	10	11 Last Quarter	12	13
14	15	16	17	18	19	20 New Moon
21	22	23	24	25	26	27 1st Quarter
28	29	30	31			

The waxing gibbous Moon will cross through the Pleiades again on December 3.

On December 6 and 7, you can find the Moon near Jupiter all night long. Interestingly, the two will be separated by roughly 7 and 6 degrees on each night, respectively.

On December 9, the waning gibbous Moon will pass extremely close to Regulus in Leo. In the far northern parts of Canada, the Moon will occult the star.

On December 26, the waxing crescent Moon and Saturn will be a mere 2 degrees apart in the western sky with faint Neptune just above them.

And finally on December 30 and 31, the waxing crescent Moon will hop past the Pleiades for a second time this month.

We get a full Moon on December 4, and last quarter falls on December 11. The new Moon is on December 20 and first quarter on December 27. The Moon will be at perigee on December 4 and at apogee on December 17.

The Moon with Jupiter below

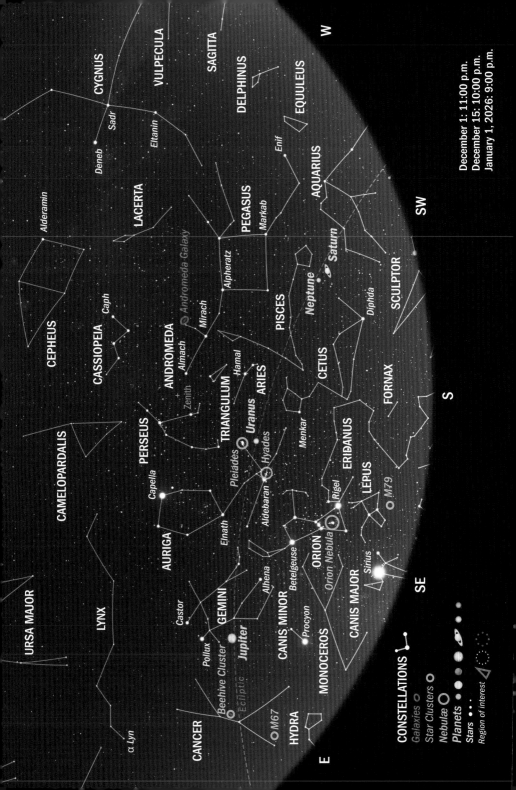

Orion Nebula (Messier 42)

Highlights in the Southern Sky

Looking southeast in evening, there is the somewhat faint – but large – constellation **Eridanus (the River Eridanus)**. **Cetus (the Whale)** has now moved to the southwest, along with **Pisces (the Fishes)** and **Pegasus (the Flying Horse)**. **Saturn** is also beginning to drop lower in the southwest.

Lepus (the Hare) can be found very low in the south. If you have a good view, you may try to use good binoculars or a small telescope to check out **Messier 79**, a globular cluster that lies roughly 41,000 light-years away.

Orion (the Hunter) has now risen nicely in the southeast, with its impressive **Orion Nebula (Messier 42)** clearly visible in his sword hanging beneath his three belt stars. Binoculars reveal the nebula as a faint fuzzy patch, even in light-polluted cities, though it's always best to try to get to a dark-sky location for the best view.

Orion's stunning red star **Betelgeuse** is unmistakable as his left shoulder (as seen from our vantage point), with rival star **Rigel** representing his right foot. The color contrast of these two stars is quite noticeable.

Also notable is **Canis Minor (the Little Dog)**, with its brightest star, **Procyon**. And, speaking of dogs, you may have noticed another extremely bright star below Orion. That would be **Sirius**, also known as the **Dog Star**, which twinkles in **Canis Major (the Great Dog)**.

Cancer (the Crab) is now climbing the eastern sky. Be sure to use binoculars to seek out the **Beehive Cluster (Messier 44)**, an extremely bright open cluster. There is another fainter open cluster in Cancer, **Messier 67**. It is sometimes called the **King Cobra Cluster**.

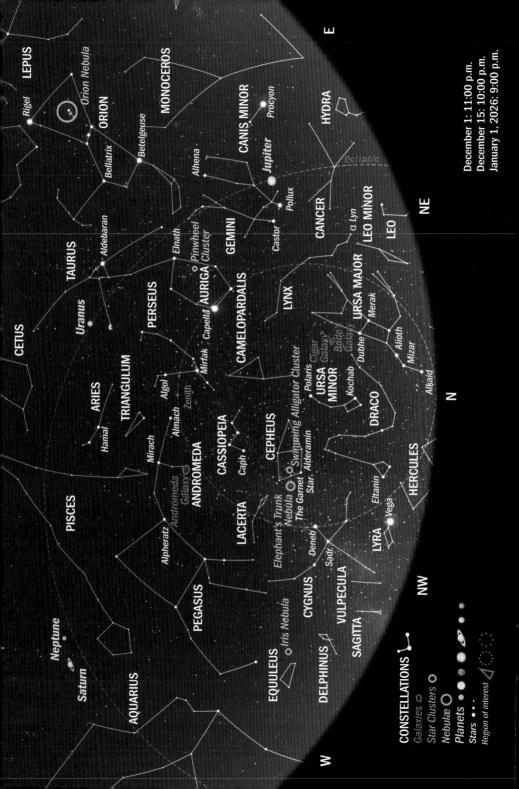

Cigar Galaxy (Messier 82)

Highlights in the Northern Sky

Cepheus (the King) is now back in the northwest, with its familiar asterism looking like a sideways house. You can see the open **Swimming Alligator Cluster (NGC 7160)** within its square, in between the stars **Alderamin** and **Zeta Cephei**. It's also a good time for astronomers to view the **Elephant's Trunk Nebula (IC 1396)** as well as the **Iris Nebula (NGC 7043)**. And let's not forget Cepheus's famous **Garnet Star (Mu [μ] Cephei)**.

Draco (the Dragon) is below Polaris in the north, with its head at the northwestern horizon.

Meanwhile, **Cassiopeia (the Queen)** can be seen high in the northwest as an "M," with **Camelopardalis (the Giraffe)** to its right. Both are just above **Ursa Minor (the Little Bear)**.

Looking up toward the zenith, you can see part of the winter Milky Way running through **Perseus (the Hero)** and **Auriga (the Charioteer)**. Binoculars will reveal several rich areas of stars and star clusters, including the **Pinwheel Cluster (Messier 36)**.

Ursa Major (the Great Bear) is climbing in the northeast, which makes this a great time to see both **Bode's Galaxy (Messier 81)** and the **Cigar Galaxy (Messier 82)**.

The Eagle Nebula (Messier 16) is a beautiful
star-forming region found in the constellation Serpens

The Messier Catalog

Name	Traditional Name	Type	Constellation
Messier 1 (NGC 1952)	Crab Nebula	Supernova remnant	Taurus
Messier 2 (NGC 7089)		Globular cluster	Aquarius
Messier 3 (NGC 5272)		Globular cluster	Canes Venatici
Messier 4 (NGC 6121)		Globular cluster	Scorpius
Messier 5 (NGC 5904)		Globular cluster	Serpens
Messier 6 (NGC 6405)	Butterfly Cluster	Open cluster	Scorpius
Messier 7 (NGC 6475)	Ptolemy Cluster	Open cluster	Scorpius
Messier 8 (NGC 6523)	Lagoon Nebula	Emission nebula with cluster	Sagittarius
Messier 9 (NGC 6333)		Globular cluster	Ophiuchus
Messier 10 (NGC 6254)		Globular cluster	Ophiuchus
Messier 11 (NGC 6705)	Wild Duck Cluster	Open cluster	Scutum
Messier 12 (NGC 6218)	Gumball Globular	Globular cluster	Ophiuchus
Messier 13 (NGC 6205)	Hercules	Globular cluster	Hercules
Messier 14 (NGC 6402)		Globular cluster	Ophiuchus
Messier 15 (NGC 7078)	Great Pegasus Cluster	Globular cluster	Pegasus
Messier 16 (NGC 6611)	Eagle Nebula	Emission nebula with cluster	Serpens
Messier 17 (NGC 6618)	Omega Nebula	Emission nebula with cluster	Sagittarius
Messier 18 (NGC 6613)		Open cluster	Sagittarius
Messier 19 (NGC 6273)		Globular cluster	Ophiuchus
Messier 20 (NGC 6514)	Trifid Nebula	Emission, reflection and dark nebula with cluster	Sagittarius
Messier 21 (NGC 6531)		Open cluster	Sagittarius
Messier 22 (NGC 6656)	Sagittarius Cluster	Globular cluster	Sagittarius
Messier 23 (NGC 6494)		Open cluster	Sagittarius
Messier 24 (IC 4715)	Sagittarius Star Cloud	Milky Way star cloud	Sagittarius
Messier 25 (IC 4725)		Open cluster	Sagittarius
Messier 26 (NGC 6694)		Open cluster	Scutum
Messier 27 (NGC 6853)	Dumbbell Nebula	Planetary nebula	Vulpecula
Messier 28 (NGC 6626)		Globular cluster	Sagittarius
Messier 29 (NGC 6913)		Open cluster	Cygnus

Name	Traditional Name	Type	Constellation
Messier 30 (NGC 7099)		Globular cluster	Capricornus
Messier 31 (NGC 224)	Andromeda Galaxy	Spiral galaxy	Andromeda
Messier 32 (NGC 221)	Le Gentil	Dwarf elliptical galaxy	Andromeda
Messier 33 (NGC 598)	Triangulum Galaxy	Spiral galaxy	Triangulum
Messier 34 (NGC 1039)		Open cluster	Perseus
Messier 35 (NGC 2168)		Open cluster	Gemini
Messier 36 (NGC 1960)	Pinwheel Cluster	Open cluster	Auriga
Messier 37 (NGC 2099)		Open cluster	Auriga
Messier 38 (NGC 1912)	Starfish Cluster	Open cluster	Auriga
Messier 39 (NGC 7092)		Open cluster	Cygnus
Messier 40	Winnecke 4	Double star	Ursa Major
Messier 41 (NGC 2287)		Open cluster	Canis Major
Messier 42 (NGC 1976)	Orion Nebula	Emission-reflection nebula	Orion
Messier 43 (NGC 1982)	De Mairan's Nebula	Emission-reflection nebula	Orion
Messier 44 (NGC 2632)	Beehive Cluster	Open cluster	Cancer
Messier 45	Pleiades	Open cluster	Taurus
Messier 46 (NGC 2437)		Open cluster	Puppis
Messier 47 (NGC 2422)		Open cluster	Puppis
Messier 48 (NGC 2548)		Open cluster	Hydra
Messier 49 (NGC 4472)		Elliptical galaxy	Virgo
Messier 50 (NGC 2323)	Heart-Shaped Cluster	Open cluster	Monoceros
Messier 51 (NGC 5194, NGC 5195)	Whirlpool Galaxy	Spiral galaxy	Canes Venatici
Messier 52 (NGC 7654)		Open cluster	Cassiopeia
Messier 53 (NGC 5024)		Globular cluster	Coma Berenices
Messier 54 (NGC 6715)		Globular cluster	Sagittarius
Messier 55 (NGC 6809)	Summer Rose Star	Globular cluster	Sagittarius
Messier 56 (NGC 6779)		Globular cluster	Lyra
Messier 57 (NGC 6720)	Ring Nebula	Planetary nebula	Lyra
Messier 58 (NGC 4579)		Barred spiral galaxy	Virgo
Messier 59 (NGC 4621)		Elliptical galaxy	Virgo

Name	Traditional Name	Type	Constellation
Messier 60 (NGC 4649)		Elliptical galaxy	Virgo
Messier 61 (NGC 4303)		Spiral galaxy	Virgo
Messier 62 (NGC 6266)		Globular cluster	Ophiuchus
Messier 63 (NGC 5055)	Sunflower Galaxy	Spiral galaxy	Canes Venatici
Messier 64 (NGC 4826)	Black Eye Galaxy	Spiral galaxy	Coma Berenices
Messier 65 (NGC 3623)		Barred spiral galaxy	Leo
Messier 66 (NGC 3627)		Barred spiral galaxy	Leo
Messier 67 (NGC 2682)	King Cobra Cluster	Open cluster	Cancer
Messier 68 (NGC 4590)		Globular cluster	Hydra
Messier 69 (NGC 6637)		Globular cluster	Sagittarius
Messier 70 (NGC 6681)		Globular cluster	Sagittarius
Messier 71 (NGC 6838)		Globular cluster	Sagittarius
Messier 72 (NGC 6981)		Globular cluster	Aquarius
Messier 73 (NGC 6994)		Asterism	Aquarius
Messier 74 (NGC 628)	Phantom Galaxy	Spiral galaxy	Pisces
Messier 75 (NGC 6864)		Globular cluster	Sagittarius
Messier 76 (NGC 650, NGC 651)	Little Dumbbell Nebula	Planetary nebula	Perseus
Messier 77 (NGC 1068)	Cetus A	Spiral galaxy	Cetus
Messier 78 (NGC 2068)		Reflection nebula	Orion
Messier 79 (NGC 1904)		Globular cluster	Lepus
Messier 80 (NGC 6093)		Globular cluster	Scorpius
Messier 81 (NGC 3031)	Bode's Galaxy	Spiral galaxy	Ursa Major
Messier 82 (NGC 3034)	Cigar Galaxy	Starburst irregular galaxy	Ursa Major
Messier 83 (NGC 5236)	Southern Pinwheel Galaxy	Barred spiral galaxy	Hydra
Messier 84 (NGC 4374)		Lenticular or elliptical galaxy	Virgo
Messier 85 (NGC 4382)		Lenticular or elliptical galaxy	Coma Berenices
Messier 86 (NGC 4406)		Lenticular or elliptical galaxy	Virgo
Messier 87 (NGC 4486)	Virgo A	Elliptical galaxy	Virgo

Name	Traditional Name	Type	Constellation
Messier 88 (NGC 4501)		Spiral galaxy	Coma Berenices
Messier 89 (NGC 4552)		Elliptical galaxy	Virgo
Messier 90 (NGC 4569)		Spiral galaxy	Virgo
Messier 91 (NGC 4548)		Barred spiral galaxy	Coma Berenices
Messier 92 (NGC 6341)		Globular cluster	Hercules
Messier 93 (NGC 2447)		Open cluster	Puppis
Messier 94 (NGC 4736)	Cat's Eye Galaxy	Spiral galaxy	Canes Venatici
Messier 95 (NGC 3351)		Barred spiral galaxy	Leo
Messier 96 (NGC 3368)		Spiral galaxy	Leo
Messier 97 (NGC 3587)	Owl Nebula	Planetary nebula	Ursa Major
Messier 98 (NGC 4192)		Spiral galaxy	Coma Berenices
Messier 99 (NGC 4254)	Coma Pinwheel	Spiral galaxy	Coma Berenices
Messier 100 (NGC 4321)		Spiral galaxy	Coma Berenices
Messier 101 (NGC 5457)	Pinwheel Galaxy	Spiral galaxy	Ursa Major
Messier 102 (NGC 5866)	Spindle Galaxy	Lenticular galaxy	Draco
Messier 103 (NGC 581)		Open cluster	Cassiopeia
Messier 104 (NGC 4594)	Sombrero Galaxy	Spiral galaxy	Virgo
Messier 105 (NGC 3379)		Elliptical galaxy	Leo
Messier 106 (NGC 4258)		Spiral galaxy	Canes Venatici
Messier 107 (NGC 6171)		Globular cluster	Ophiuchus
Messier 108 (NGC 3556)	Surfboard Galaxy	Barred spiral galaxy	Ursa Major
Messier 109 (NGC 3992)		Barred spiral galaxy	Ursa Major
Messier 110 (NGC 205)	Edward Young Star	Dwarf elliptical galaxy	Andromeda

The galaxies Messier 95 (bottom) and
Messier 96 (top) can both be found in Leo

Glossary

annular solar eclipse: A kind of partial solar eclipse during which the Moon does not completely cover the Sun's disk but leaves a ring of sunlight around the Moon.

aphelion: For an object orbiting the Sun, the point in its orbit that is farthest from the Sun.

apogee: When the Moon (or any satellite of Earth) is at its most distant in its monthly orbit around Earth.

arcsecond (or second of arc): A tiny angle equal to 1/3600 of a degree, used to measure the separation of double stars and the apparent diameters of Solar System objects.

asterism: A group of stars within a constellation (or sometimes from more than one constellation) that forms its own distinct pattern.

astronomical unit (au): A unit of distance that uses the average distance between Earth and the Sun as its base metric. One astronomical unit is approximately 150 million kilometers (93.2 million miles).

autumnal (fall) equinox: The date in September marking the end of summer and the beginning of autumn, when the Sun illuminates the Northern and Southern Hemispheres equally. The lengths of the day and night are equal, and it's one of two instances of the year that the Sun has a declination of 0 degrees.

binary star system: A double-star system in which two stars are gravitationally bound to each other and orbit a common center of mass. Many double-star systems have multiple components.

conjunction: Technically speaking a conjunction is when two objects have the same right ascension, but amateur astronomers also use the term when there is a close approach of two or more celestial objects.

constellation: A group of stars that make an imaginary image in the night sky. The International Astronomical Union (IAU) has designated the boundaries between 88 official constellations covering the celestial sphere.

degree: A unit of measurement used to measure the distance between objects or the position of objects in astronomy. The entire sky spans 360 degrees, and up to about 180 degrees of sky is visible from any given point on Earth with an unobstructed horizon.

double star (binary star): A pair of stars with small angular separation (from fractions of an arcsecond to hundreds of arcseconds). Optical doubles are chance alignments of stars with no physical connection to each other. True binary stars are gravitationally bound (see binary star system).

fireball: A very bright meteor, generally brighter than magnitude –4.0 (about the brightness of Venus).

galaxy: An enormous system of gas, dust and billions of stars and their planetary systems, all held together by gravity.

globular cluster: Old star systems at the edges of spiral galaxies that can contain anywhere from thousands to millions of stars, packed in a close, roughly spherical form and held together by gravity.

greatest eastern elongation: When an inner planet (Mercury or Venus) is farthest from the Sun in the evening sky.

greatest western elongation: When an inner planet (Mercury or Venus) is farthest from the Sun in the morning sky.

light-year: A unit of distance that uses how far light travels in one Earth year as its base metric. One light-year is about 9 trillion kilometers (6 trillion miles). Objects in outer space are so far apart that it takes a long time for their light to reach Earth, and so the farther an object is, the farther in the past that observers on Earth are seeing it. As an example, the Andromeda Galaxy (Messier 31) is 2.5 million light-years away, so when observers look at it in the sky, they are seeing it as it appeared 2.5 million years ago.

magnitude: The apparent brightness of an object in the sky. Magnitude is measured on a scale where the higher the number, the fainter the object appears. Objects with negative numbers are brighter than those with positive numbers.

meteor shower: An atmospheric event during which a number of meteors can be seen radiating from one point in the sky. Meteor showers occur when Earth, at a particular point in its orbit, crosses a stream of particles left over from a passing comet or asteroid.

meteor shower peak: The best time to view a meteor shower, when meteor activity will be at its highest.

nebula: A cloud of dust and gas visible in the night sky either as a bright patch or a dark shadow against other luminous matter. To learn more about the different types of nebulae, see page 42.

occultation: When one object appears to pass in front of another more distant object, obscuring the observer's view. For example, a lunar occultation is when the Moon appears to pass in front of a more distant object, like a planet or star.

open cluster: A star system that contains anything from a dozen to hundreds of stars, in which the stars are spread out. Open clusters are mostly found near the galactic plane, the plane on which most of a galaxy's mass lies.

opposition: When an outer planet (Mars, Jupiter, Saturn, Uranus or Neptune), a dwarf planet or a minor planet is opposite the Sun in the sky in right ascension.

partial lunar eclipse: A lunar eclipse during which the Earth's shadow partially covers the Moon.

partial solar eclipse: A solar eclipse during which the Moon only partially covers the Sun.

perigee: When the Moon (or any satellite of Earth) is at its closest in its monthly orbit around Earth.

perihelion: For an object orbiting the Sun, the point in its orbit that is nearest to the Sun.

star: A luminous ball of plasma held together by its own gravity and fueled initially by the nuclear fusion of hydrogen into helium at its core. Stars come in a vast variety of sizes, luminosities and temperatures, typically ranging from dwarf sizes (that are as little as 10 percent the mass of the Sun) to hypergiants (that can be 100 or more times the mass of the Sun). To learn more about the different types of stars, see pages 14–15.

summer solstice: The point during the year that a particular hemisphere is most tilted toward the Sun. This takes place in June in the Northern Hemisphere.

supernova: A powerful and luminous explosion that occurs during the last evolutionary stages of a massive star when its brightness increases immensely and it throws off most of its mass; a supernova can also be when a white dwarf is triggered into runaway nuclear fusion by the accretion of material from a binary companion; or by a stellar merger.

total lunar eclipse: A lunar eclipse during which the Earth's shadow completely covers the Moon. Such an eclipse can last for hours.

total solar eclipse: A solar eclipse during which the Moon covers the entire face of the Sun, revealing the Sun's corona and prominences. Solar eclipses can last from a few seconds to about seven minutes maximum at a specific location.

variable star: A star whose apparent magnitude varies over time. The variability might be caused by physical changes to the star itself or by how it rotates or interacts with nearby objects.

vernal (spring) equinox: The date in March marking the end of winter and the beginning of spring, when the Sun illuminates the Northern and Southern Hemispheres equally. The lengths of the day and night are equal, and it's one of two instances of the year that the Sun has a declination of 0 degrees

winter solstice: The point during the year that a particular hemisphere is most tilted away from the Sun. This takes place in December in the Northern Hemisphere.

Zenithal Hourly Rate (ZHR): The rate of meteors a shower would produce per hour under clear, dark skies and with the radiant at the zenith.

The Greek Alphabet (lowercase)

α	alpha	ζ	zeta	λ	lambda	π	pi	φ	phi
β	beta	η	eta	μ	mu	ρ	rho	χ	chi
γ	gamma	θ	theta	ν	nu	σ	sigma	ψ	psi
δ	delta	ι	iota	ξ	xi	τ	tau	ω	omega
ε	epsilon	κ	kappa	o	omicron	υ	upsilon		

Resources

American Astronomical Society — aas.org — An organization of professional astronomers seeking to enhance and share humanity's understanding of the Universe

American Meteor Society — amsmeteors.org — A non-profit scientific organization that informs, encourages and supports research in Meteor Astronomy

Eclipsewise.com — eclipsewise.com — Astronomer Fred Espenak's site dedicated to predictions and information on eclipses of the Sun and Moon

European Southern Observatory — eso.org — An intergovernmental research organization for ground-based astronomy that shares recent research, news and images

European Space Agency (ESA) — esa.int — The main site for the European Space Agency that shares news, research and launches

Hubblesite.org — hubblesite.org — NASA's main site for the Hubble Space Telescope, including news and images

James Webb Space Telescope — webb.nasa.gov — NASA's main site for the James Webb Space Telescope, including the latest images

NASA — nasa.gov — An update on all NASA missions, including astronomical events, news and launches

The Planetary Society — planetary.org — An international, non-profit organization that promotes the exploration of space through education, advocacy and research

The Royal Astronomical Society of Canada — rasc.ca — Home to Canada's national astronomical association

Spaceweather.com — spaceweather.com — An update on space weather, including solar flares, coronal mass ejections and noctilucent cloud forecasts

Space Weather Prediction Center — swpc.noaa.gov — The National Oceanic and Atmospheric Administration's site on forecasts for space weather, including potential geomagnetic storms

Solar and Heliospheric Observatory — soho.nascom.nasa.gov — A collaborative project between the ESA and NASA to study the Sun and its interactions with Earth from different perspectives

Solar Dynamics Observatory — sdo.gsfc.nasa.gov — Satellite observations of the Sun

Photo Credits